Anahi Chimini Sobral
Maria C. Perusi
Victor F. Nascimento

Impacts of irregular urban land use

Anahi Chimini Sobral
Maria C. Perusi
Victor F. Nascimento

Impacts of irregular urban land use

Soil degradation due to erosion and alteration of the soil's physical properties in a watershed in the municipality of Ourinhos/SP

ScienciaScripts

Imprint

Cover image: www.ingimage.com

This book is a translation from the original published under ISBN 978-3-330-77053-9.

Publisher:
Sciencia Scripts
is a trademark of
Dodo Books Indian Ocean Ltd. and OmniScriptum S.R.L publishing group

120 High Road, East Finchley, London, N2 9ED, United Kingdom
Str. Armeneasca 28/1, office 1, Chisinau MD-2012, Republic of Moldova, Europe
Managing Directors: Ieva Konstantinova, Victoria Ursu
info@omniscriptum.com

Printed at: see last page
ISBN: 978-620-8-64778-0

Summary

This book is the result of Anahi Chimini Sobral's Bachelor's Degree in Geography, presented in 2011 to an examining board at the Universidade Estadual Paulista, UNESP, Ourinhos/SP Campus.

1 Introduction

The process of urban expansion without proper planning and in inappropriate locations, such as steep slopes and spring areas, can result in the degradation of natural resources, especially vegetation cover, soil and water. Notorious examples include the establishment of allotments in areas that should be permanently preserved, the channeling of rainwater to inappropriate locations and the intense process of soil sealing. One of the consequences of these procedures is the establishment of erosion, known as urban erosion. In this sense, it is important to draw up public and private works projects that balance urban development and environmental conservation, so that the expansion of the urban network has a less negative impact on natural environments.

Soil is one of the natural resources most affected by the urbanization process, paving the way for a discussion about urban soils, as well as their use and occupation. According to Pedron et al. (2007) these soils have an urban environment over them, and this classification is independent of the degree of alteration the soil is undergoing. In this sense, soils in an urban environment tend to undergo drastic changes in their morphological, physical, chemical and biological characteristics, making them totally different from those which, even in an urban environment, have not undergone any changes (SCHARENBROCH et al., 2005, cited by DALMOLIN, 2006).

In this context, the aim was to study an area degraded by water erosion on the banks of Avenida Vitalina Marcusso, in the municipality of Ourinhos/SP (Figure 1). The general objective is based on the study of eroded soils, in order to understand the physical changes suffered by the soil. The aim was also to understand the dynamics of urban erosion; to this end, maps were drawn up of land use and occupation in the Agua da Veada stream micro-basin over different periods.

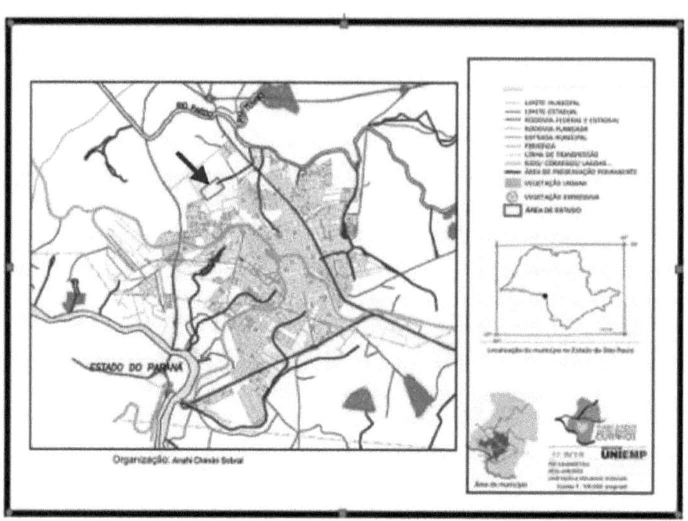

Figure 1: Location of the municipality of Ourinhos in the southwest of the state of São Paulo and indication of the study area by the blue arrow and red square.

Modified by: Sobral (2008)

The site underwent a "recovery" process funded by the Growth Acceleration Plan (PAC). The degraded area has been modified over the years due to human intervention in its natural dynamics, which has led to the formation of a large erosion site that has damaged the environment, as well as the economy and the quality of life of nearby residents. This is based on the accumulation of garbage, which can lead to the spread of disease. Ourinhos City Hall has implemented the practice of contour lines on the site, with the aim of slowing down the speed of water flowing over the land, thus reducing its erosive capacity. The erosion was filled in with leftover building materials, but the proper management of the practice was not applied, i.e. the technique was not followed up and/or monitored, as the structure is undergoing a process of slumping of the layers. Therefore, due to the picture presented and the history of alterations suffered by the area, a study of the site is considered pertinent.

Therefore, the general objective of this work was to physically characterize the soils degraded by water erosion located on Avenida Vitalina Marcusso, in the municipality of Ourinhos/SP, and to understand the space-time evolution of the erosion process.

2 Soils: concept and main physical characteristics

The natural resource soil is extremely important for the activities carried out by human beings and can be conceptualized in different ways depending on how it is used. According to Freire (2006), people use the concept of soil according to different perspectives and interests. Freire (2006, p. 1, 7) defines soil as:

[...] a natural substrate of terrestrial ecosystems, being fundamentally important for the maintenance of plant and animal communities, and its functional efficiency constitutes one of the supports of human life on Earth, the security of nations and the stability of societies.

...a natural body, made up of mineral and organic matter, which has horizons developed as a result of a particular interaction of soil-forming factors, which covers, in a more or less continuous way, most of the continental surface of the Earth, which constitutes the natural substrate of life for most higher plants, especially cultivated plants.

This natural resource is formed by the action of weathering agents that physically and chemically modify rocks and their minerals when exposed to the environment. It is a natural process of accumulation and evolution of sediments, which are deposited on the most varied forms of relief. Physical weathering disintegrates rocks, while chemical weathering alters their composition and minerals. It is therefore on this altered material, called parent material, that soils are formed. The presence of decomposed organic material is a priority for soil formation. The main agents involved in soil formation are: the parent material, climate, biological activity of living organisms, topography and time (VIEIRA, 1988; BERTONI; LOMBARDI NETO, 2005). Bertoni and Lombardi Neto (2005, p. 37) state:

The climate, represented by rainfall and temperature, mainly influences the varied distribution of soluble elements and the speed of chemical reactions. The main action of soil micro-organisms is to break down plant remains. Topography is influenced by the transverse and lateral movement of water. The formation of a soil depends, of course, on the length of time in which the different factors act.

According to Palmieri and Larach (2004, p. 66), soil is made up of three-dimensional natural bodies, which result from the interaction between climate and organisms, affecting the source material of the rock, configuring a relief that presents characteristics that express relevant processes and mechanisms in the formation of this natural resource. Thus, soil is the result of the actions of agents responsible for weathering, and its formation takes place through elements that make up a natural process marked by the action of climate, biosphere, parent rock, relief and time (VIEIRA, 1988). Soil also comprises the substrate that serves as the basis for animal and plant interactions on the earth's surface, serving as a source of food generation, profits, the basis for housing and civil construction, as well as being an essential element in the process of speculation by some landowners. Due to the many transformations to which societies are subjected, the soil element is also transformed, taking on anthropic characteristics depending on man's interaction, for example, the evolution of cities, lack of conservation in rural or urban areas, among others. According to EMBRAPA (1999, p. 5), soils are:

A collection of natural bodies, made up of solid, liquid and gaseous parts, three-dimensional, dynamic, made up of minerals and organics, which occupy most of the surface mantle of the continental extensions of our planet, contain living matter and can be vegetated in nature, where they occur. Occasionally they may have been modified by human activities.

Kiehl (1979, p. 15) uses the concept of soil to characterize it as "a system composed of three phases: solid, liquid and gaseous". The solid phase, or dispersed phase, is predominant and made up of organic and inorganic matter, constituting the soil matrix; the liquid phase, or dispersed phase, is the soil solution, made up of water, salts and suspended colloidal particles; the gaseous phase is the soil air, which differs from atmospheric air due to the percentage proportion of its components. This system described above is closely related to the main physical properties of the soil, such as color, texture, structure, consistency, density of the soil, particle density, porosity and waxyness, which will be explained below.

- Color:

According to Resende et al. (2007, p. 59), "color can provide indications about the source material, organic matter content, drainage conditions and Fe oxide content (P fixation), giving an idea about the general fertility of the soil". Freire (2006) states that by analyzing the color of the soil in the field, inferences can be made about its behavior. The different shades of soil also make it possible to differentiate horizons in field work, and during the morphological description activity this property is expressed in accordance with the Munsell Color Chart, which according to Lepsch (2002, p. 27):

The color must be described by comparison with a standardized scale. The most widely used is the "Munsell table", which consists of about 170 small rectangles with different colors, arranged systematically in a book with detachable pages. Soil color notation is done by comparing a fragment or log of a given horizon with these rectangles. Once the closest color is found, the three basic elements that make up a given color are noted: **hue** - "pure color" or fundamental rainbow color, determined by the wavelengths of light reflected from the sample. **Value** - measure of the degree of light clarity or shades of gray present (between white and black) ranging from 0 (for absolute black) to 10 (for pure white). **Chroma** - the ratio of the mixture of the fundamental color and the shade of gray, also ranging from 0 to 10.

- Texture:

The texture is given by the solid part of the soil, which can have a coarse or clayey appearance, depending on the characteristics of the particles that make it up. According to Resende et al. (2007), this solid or inorganic part is made up of particles such as sand, clay and silt, which make up the so-called fine soil, as well as larger particles such as pebbles and gravel. This author also states that "the texture or granulometric composition of a soil depends, like other attributes, on the rock of origin and the degree of weathering (age) of that rock" (RESENDE, 2007, p. 60). This means that texture expresses, in most cases, the characteristics of the parent material, and can be modified due to erosive or depositional phenomena, a fact correlated to anthropic action. Oliveira et al. (1992, p. 35) states that this property "is closely related to structure, consistency, permeability, cation exchange capacity, water retention and phosphate fixation", as well as being used as a criterion for classifying soils. As far as this research is concerned, the identification triangle used by Lepsch (2002) was used (Figure 2)

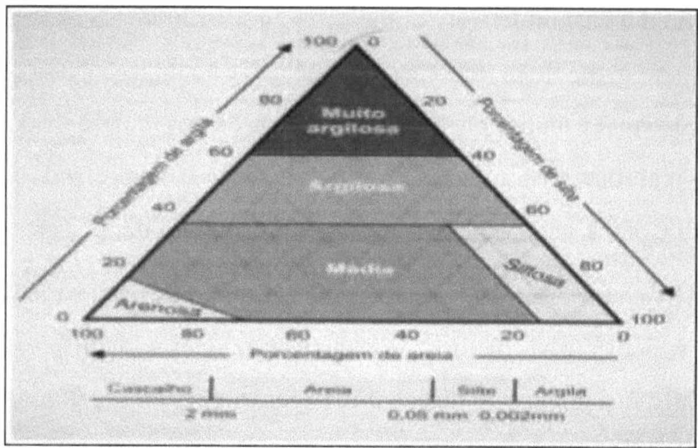

Figure 2: Triangle used to identify textural class. Source: Lepsch (2002)

- Structure:

Freire (2006, p. 19) works on the concept of structure, affirming its variability, *since* it depends on and changes according to "the season, the plant growing in the soil, its stage of development and the type of management applied to it. Structure, however, modifies the manifestation of properties that depend on texture".

According to Azevedo et al. (2006), the structure of a soil can be defined as the result of the arrangement of its primary particles, giving rise to structural units. These primary particles are sand, clay, silt, among others, such as salts, minerals and organic matter, which group together to form these units called aggregates. According to Resende et al. (2007), if the soil mass does not have a defined structure, it is characterized as massive or as loose grains (presence of non-stable aggregates).

Chaves and Calegari (2001) work with the concept that soil aggregation is based on the union of particles (clay-ion-organic matter and silt) in secondary units, with the stability of the aggregates being characterized as resistance to degrading action, particularly from water. This aggregation controls the internal movement of water, air, heat and the proliferation of roots. The addition of plant residues, organic compounds and the mowing of invasive plants increases the aggregation of soil particles and is recommended for depleted soils.

Silva and Mielniczuk (1997) already assume that the formation and stabilization of soil aggregates occurs through the action of physical, chemical and biological processes, which have their own mechanisms of action, involving substances that act on soil aggregation and stabilization. In addition to aggregating substances, there are aggregating agents, represented by climate, roots, microorganisms and others.

- **Consistency:**

Kiehl (1979) explains that the consistency of a soil depends on its moisture content, i.e. it varies according to the soil's state of moisture, which can be dry, moist or wet. Freire (2006, p. 34) states that:

Consistency is the manifestation of the forces of adhesion and cohesion that act on the soil mass as a result of variations in water content. As the concentration of soil in the soil-water system increases, the mass becomes non-flowing and the forces of adhesion and cohesion manifest themselves. These concepts mean that consistency includes various soil properties, such as compressive strength, shear strength, friability, plasticity and stickiness.

Resende et al. (2007) point to the importance of this property in relation to the management of agricultural crops on the most varied types of soil, since the identification of this property provides information on the ideal point for tillage, a limit which, if not respected, will certainly result in compaction and consequent vulnerability to water erosion.

- **Soil density**

According to Freire (2006, p. 27), "density is the ratio between the specific weight of a body and the specific weight of water under normal conditions of temperature and pressure". Bertoni and Lombardi Neto (2005) therefore state that soil density is the volume of natural soil, including the spaces occupied by air and water. It is variable and depends on structure and compaction. The less structured and compacted the soil, the higher the density value. Azevedo et. al (2006) points out that soil density decreases as the profile deepens due to the pressure of the underlying layers, causing the phenomenon of compaction, which reduces porosity and consequently restricts plant growth. This property is used to analyze the degree of compaction and also to measure changes in the structure and porosity of the soil, a fact that generally occurs in urban soils. According to Dalmolin et al. (2006), the values that can be assigned to the density of these soils are between 1.5 and 1.9 g. cm-3, and can reach 2.0 g. cm-3, a fact that marks the modification of soils and water retention capacity in the profile.

- **Particle density**

This property is characterized by the ratio between the mass of a soil sample and the volume occupied by its solid particles, disregarding the porosity factor, and does not vary according to the management applied to the soil, as it depends on the soil's chemical and mineral composition, as well as establishing the predominance of inorganic or organic components in a sample (REINERT; REICHERT, 2006). According to Freire (2006, p. 28) the particle density of Brazilian soils "varies between 2.60 and 2.75, depending on the composition of the soil. This is because the predominant minerals such as quartz, feldspars and clays have specific weights within those limits", so Kiehl (1979) admits a range of values between 2.3 and 2.9 kg.dm-3. The amount of organic matter, when abundant, can considerably lower the value of the property in question.

- **Porosity**

According to Guerra (2005), the porosity of a soil is given by the volume not filled by its solid particles, this volume being occupied by water and air, i.e. the transfer of solid particles, liquids, gases and biological activities occur in these spaces. Freire (2006) states that the proximity of the solid particles in a soil is responsible for its porosity, with constituents far apart resulting in high porosity and particles very close together resulting in low porosity in a soil. In addition, the arrangement of the solid particles also determines the size of the pores, i.e. the structure of the soil is responsible for its porosity. Selby (1985) cited by Bigarella (2007) points out that porosity is improved by the action of living organisms, plant roots or the cultivation of the land; however, it can be reduced by the effect of compaction, which can occur through the use of agricultural machinery, by the sealing of the surface caused by splashing, because the fragmented particles of the soil aggregates begin to occupy the spaces provided by the pores.

Pores are empty spaces between soil aggregates in which soil solution and air are housed, and the relative distribution of different pore sizes is a function of soil structure. Porosity can be defined as the space in the soil not occupied by the matrix (organic and inorganic components), depending directly on the texture and structure of the soil. Microporosity is the main factor responsible for water retention, while macroporosity allows water to run off fairly quickly and the voids are occupied by air. The loss of porosity is associated with a reduction in the content of organic matter, compaction and the effect of the impact of raindrops, all of which cause a reduction in the size of the aggregates and consequently reduce the size of the pores (BERTONI; LOMBARDI NETO, 2005).

- **Cerosity**

For Prado (2003, p. 17) cerosity "are films that can cover the structural aggregates of the B horizon in different quantities and degrees of clarity". Resende et al. (2007, p. 68) start from the premise that cerosity is a thin film of clay that is deposited on the lower horizons, but that it can also form from the "rearrangement of the colloidal material on the surface of the blocks or prisms, due to the alternating expansion and contraction of the soil material" and is used to indicate the relative richness of the nutrients in a soil, in addition to the fact that its presence or absence in a soil is directly linked to the capacity of the roots to absorb nutrients. Santos et al. (2005, p. 33) conceptualize it as:

[...] the somewhat shiny and waxy appearance of the natural surfaces that cover the different faces of structural units, often manifested by a more intense color, and the coated surfaces are usually free of bare grains of sand and silt.

According to Oliveira et al. (1992) its occurrence is related to the B or C horizons, as well as being an important characteristic in determining the textural B horizon.

Thus, from what was presented, it was possible to observe the characteristics of soil formation, as well as its conceptualization from the point of view of various authors. The main physical properties of soils were also discussed, including the concepts and particularities of each one. It is important to note that these characteristics can change as materials of technogenic origin are incorporated, which modify the morphology of soils.

2.1 Soil profile, horizons and layers

The soil profile is made up of vertical sections parallel to the surface, which go deeper until they reach the parent material. These sections are called horizons and layers (SANTOS et al., 2005; VIEIRA, 1988) (Figure 3).

Figure 3: Schematic of a soil profile.

Source: Virtual Teacher Reference Center

According to Santos et al. (2005, p. 2), the horizons "are the result of the action of formation processes, with a genetic relationship between them within the profile. By worldwide convention, they are represented by the letters O, H, A, E, B and C from the surface towards the parent material". The layers are little or very little worked by weathering processes, such as rock and organic matter in an unhumified stage "as a result of the intensity of life in the soil, disintegration, as well as the physical and chemical transformations carried out on it" (VIEIRA, 1988, p. 120). Guerra (2004, p. 67-68) states that:

The soil profile is the unit for describing and examining soils in their natural environment. It comprises the identified pedogenetic horizons, contained in a vertical section, from the surface of the ground to a depth of 2 meters, or until the appearance of rock in the initial stage of decomposition, or not, if this occurs at a depth of less than 2 meters (or until the appearance of a diagnostic horizon, if this is at a depth greater than 2 meters).

Kiehl (1979) works on the assumption that soil profiles are formed through physical and chemical weathering of rocks, forming altered material in its composition, but which retains characteristics of the parent rock. This material could move mechanically or be dissolved in water from the soil or rainfall, forming different horizons by removal and/or deposition.

A soil profile is made up of horizons and layers, which can be differentiated in the field using criteria such as "texture, color, consistency, structure, biological activity, type of aggregate surface, etc." (VIEIRA, 1988, p. 120). In the laboratory, samples can be subjected to physical, chemical and mineralogical analysis, so that results can be obtained about their differentiation.

A well-developed soil usually has the following horizons:

Horizon O: Streck et al. (2008, p. 18) state that this "is a horizon or superficial organic layer made up of little or undecomposed organic remains, formed in a well-drained environment or occasionally saturated with water".

Horizon H: Santos et al. (2005) state that this is an organic horizon whose material can be found in the most diverse forms of decomposition, and can be superficial or not, occurring in areas where there has been prolonged water stagnation.

Horizon A: According to Oliveira (1992), horizon A is superficial and has mineral characteristics with an accumulation of organic matter near the surface. This is the part of the soil that is most interesting for growing plants, as it is notable for the intense movement of macro and micro flora and fauna, which produce organic matter. Its color is darker than the underlying horizons and its thickness is variable, in some cases reaching a depth of around 100 cm.

E horizon: According to Streck et al. (2008), this is a mineral horizon whose color is lighter due to the loss of clay, iron oxides or organic matter, elements which have been transferred to the B horizon by eluviation. Characterized by a residual concentration of sand, it generally occurs below the A or O horizon.

Horizon B: According to Oliveira (1992), this is a mineral horizon, endowed with stable pedogenetic properties due to its location along the profile, a fact that configures it as a horizon that does not normally suffer from losses, since its transformations occur from alteration and decomposition of the source material. It is used as a diagnostic horizon when determining soil classes in Brazil.

Horizon C: Santos et al. (2005) state that these are identified as layers of sediment, saprolite, unconsolidated rock material and other uncemented rock materials that do not show strong resistance when excavated or cut with a shovel.

Layer R: unaltered rock.

2.2 Latosols

Based on the Pedological Map of the State of São Paulo, EMBRAPA (1999), red latosols predominate in the municipality of Ourinhos. Latosols are formed from mineral material and have a latosolic B horizon, immediately below any type of A horizon or surface horizon, within 2 to 3 m of the soil surface depending on the thickness of the A horizon (EMBRAPA, 2006). The texture varies from clayey to medium and the predominant color is reddish. Most of them have good physical properties, with total porosity between 50-60 % and, consequently, good internal drainage, even in clay-textured specimens (LEPSCH, 2002). Streck et al. (2008) conceptualize Latosols as deep, well-drained soils with little differentiation from one horizon to another, a high degree of weathering, which results in the presence of low-activity clay, low base retention and a relative absence of weatherable primary minerals. It has a profile with a sequence of A-Bw-C horizons, the Bw being the diagnostic horizon of latosols. These soils generally have high acidity and low nutrient reserves. According to Streck et al. (2008, p. 208), commenting on the Bw horizon or B

latosol:

It is characterized by being a B horizon with no textural gradient in relation to the A horizon, well drained, with a weak to moderate blocky or strong microgranular structure, low activity clay (<17 cmolc/kg of clay); low content of weatherable minerals (< 4%) and rock fragments (< 5%), with a sandy loam or more clayey texture.

A striking characteristic of latosols is the diffuse or gradual transition between their horizons, which results in little morphological variation. They are located on gently undulating to undulating terrain, with slopes varying between 2% and 15% (NUNES, 2003). Oliveira (1999) states that due to the high internal permeability and low adsorptive capacity of these soils, they can be characterized as poorly filtering, which means that, despite their thickness, there is a high possibility of contamination of aquifers by toxic material deposited in them.

Latosols have four suborders: brown, yellow, red and yellow-red, which are divided into large groups, which in turn are again subdivided into subgroups. In the research area, based on the Pedological Map of the State of São Paulo (EMBRAPA, 1999), there is predominantly the red suborder: "soils with a 2.5YR or more red hue in most of the first 100 cm of the B horizon (including BA)" (OLIVEIRA, 1999).

In this way, the main characteristics of Latosols have been explained, as this is the predominant soil class in the state of São Paulo (OLIVEIRA, 1999). It is important to clarify that soil horizons, as well as their physical characteristics, can be modified by anthropogenic action to the extent that materials are added to their configuration. When these materials are incorporated into the soil, they start to behave like deposits containing technogenic materials, a formation that is only possible due to human interference, a fact that modifies the landscape and provides, for example, historical information on the process of occupation and the formation of urban centers.

2.3 Erosion

One of the ways in which soils are degraded is through erosion on hillsides. The erosion process consists of the disintegration, transportation and deposition of sediments at the foot of slopes or in watercourses. It can manifest itself in various ways: laminar erosion, gullies and gullies. It is a geological phenomenon that occurs independently of human action, but its balance can be disrupted by human intervention.

In Brazil, water erosion is the most common erosion process. It is caused by rainfall runoff and its main causes are the impact of raindrops on bare ground and the resulting detachment of soil particles. Erosion is therefore the process of detachment, dragging and deposition of soil particles caused by water and wind (BERTONI; LOMBARDI NETO, 2005).

Guerra (2005) states that the start of the erosion process is mainly due to the rupture of aggregates during the *splash* action that occurs in bare soil, which depends on both the resistance of the soil to the impact of raindrops and the erosivity of the rain. The broken aggregates are easily carried away by surface runoff, which intensifies as soon as the soil is sealed, hindering the infiltration of rainwater and forming puddles that break off and start the runoff process. According to Coelho Neto (2003, p. 119), "the process of infiltration results from the interdependent relationships between the mechanisms of entry into the soil surface, storage within the soil and transmission of soil moisture". Due to irregularities in the topography, linear flows are formed that concentrate runoff, causing the formation of micro-ravines, which can develop into the most advanced stage of degradation: gullies. However, according to Salomao (2005), the formation of gullies is mainly related to the combination of natural factors, such as the susceptibility of the soil to eroding, and the use and occupation to which the soil is subjected. The author states that gullies occur when surface and subsurface water act together, making them "the scene of various phenomena: surface erosion, internal erosion, undermining, landslides and slides, which combine to give this form of erosion great destructive power" (SALOMÂO, 2005, p. 230).

According to Perusi (2004), it occurs due to the imbalance of the water/vegetation/soil system, starting with the impact of raindrops on soil aggregates, breaking them up and making them vulnerable to surface runoff, which can form runoff channels that can develop into gullies and gullies, It is also responsible for removing colloidal particles from the soil, i.e. clay and organic matter, which play an important role in transporting nutrients, water and mineral salts, damaging plant growth as soil fertility is reduced.

When it occurs in urban areas, this deflagrating degradation of the soil is known as urban erosion, which is associated with the inefficiency and inadequacy of urban planning and development policies with regard to land-use practices for construction. Ramos (1995) cited by Silva et al. (2003-2004, p.

48) states that:

Erosion and surface sediment transport in an urbanized region differ significantly from the processes that occur in regions where land use is predominantly for rural purposes. In urban areas, although similar types of erosion can occur to rural areas, those resulting from the concentration of flow predominate, mainly due to deficiencies in the drainage system.

Still on the subject of urban erosion, Salomao (2005) states that in addition to representing areas susceptible to accidents, they are mistakenly likened to deposits of urban waste and sewage. Often this deposit comes with the intention of controlling its evolution, but this practice has failed, as it transforms the erosion into a focus for disease, further aggravating the environmental damage caused by it.

2.4 Urban soils

According to Mota (1999), the urban environment is made up of a "natural system", made up of the physical and biological environment, and an "anthropic system", which corresponds to man and his activities, which are intrinsically related. Corrêa (1989), quoted by Rocha and Foresti (1998), states that land use is a reflection of the socio-economic relationships established in a territory and, in this way, demonstrates the appropriation of nature by man and the changes imposed on it. Based on this assumption, it can be said that in the city of Ourinhos/SP, the process of appropriation of the territory and its consequent urbanization often takes place inappropriately, resulting in the degradation and destruction of peripheral areas whose presence of springs and streams, due to these characteristics, calls for the preservation of the environment.

According to Pedron et al. (2007, p.147), soil, formed from the interaction of various factors, is one of the natural resources most affected by urbanization. The authors add:

[...] soil has vital functions for the urban system, such as supporting engineering works and plant life, as well as acting as a natural filter, regulating the hydrological cycle and preventing various toxic substances from being dispersed into the environment. Thus, most of the activities resulting from the urbanization process will directly affect the soil resource, with greater or lesser intensity, and can often increase the degree of environmental degradation, also affecting the population's quality of life.

From this point on, it is possible to call them urban soils, which, according to Pedron et al. (2007), are nothing more than soils found in the urban environment that may or may not have undergone changes due to human action. For soils that have undergone changes due to human action, the aforementioned author recommends the use of the nomenclature anthropic soils, therefore configured as those "significantly modified by the intense and continued use of man through agricultural, mineral and urban exploitation, among others" (PEDRON et al., 2007). Gomes et al. (2008) define anthropic soils or Anthroposols as those that show dense human intervention through the incorporation of inactive materials and/or the partial removal of soil, which must be at least 40 centimeters deep as a response to this intervention.

Santos Jùnior (2008, p. 04) states that urban soils can be defined as:

[...] unconsolidated mineral and/or organic material on the Earth's surface, altered either by mixing, deposition or contamination by various materials caused by human activities in urban or urbanized areas and not related to agricultural production, capable of supporting the development of plants (CRAUL, 1992; BOCKHEIM, 1974 and FANNING; STEIN; PATTERSON, 1978).

According to Martins (s.a), these soils, which are directly in contact with each other and serve as a support for the growth of cities, must be protected by conservation policies and measures that prevent or at least minimize their degradation, as well as actions that promote the rational use of their potential. It is in this context of constant urbanization and degradation of the natural resource soil that the area studied is inserted. It can therefore be said that the situation of degradation found today is related to the creation of a disorganized territory in terms of management and planning. Nefussi and Licco (2005, p. 1) state that:

Since the 1950s, the formation of Brazilian cities has created a contrasting scenario, typical of the big cities of the Third World. The way in which the majority of municipalities were created ended up trampling on the models of territorial organization and urban management traditionally used, and proved to be inadequate. The result has been the emergence of cities without the infrastructure and availability of urban services to cope with the growth caused by the population that has migrated to the cities.

Based on this statement, it can be seen that the urban environment encompasses not only social and economic problems, but also the reflection of the lack of planning and management of basic infrastructure, such as sanitation, which is precarious in many municipalities, which also causes environmental problems, so social issues are intertwined with environmental issues. Nefussi and Licco (2005, p. 2) add:

When it comes to cities, the complexity of what are known as environmental problems requires special, transdisciplinary treatment. Cities are not just spaces where social problems are evident. The built environment itself plays a preponderant role in constituting the problem, which transcends the physical environment and involves cultural, economic and historical issues.

Nefussi and Licco (2005) conclude that the expansion of cities and the consequent misuse of soil in urban environments is responsible for the degradation of the natural resource in question. This degradation alludes to the various modifications that soils undergo, which have implications for the urban environment due to the alterations suffered by the physical and chemical properties of soils.

One of these changes, according to Schleuβ et al. (1998), cited by Pedron et al. (2004), is morphological and refers to alterations that occur due to the removal of the topsoil and/or the addition of materials in order to reconstitute the lost soil. Under these conditions, the soil becomes heterogeneous, since different materials with different dimensions and textures have been introduced to it, a fact that interferes with the soil's hydric and thermal regime, also affecting its ability to support plants and its resistance to erosion and landslides.

Dias Júnior (2000) points out that soil compaction is a phenomenon related to a reduction in volume and consequent increase in soil density, since compaction reduces pore size.

It is therefore understood that soil compaction has a detrimental effect on soil porosity, which will guide land use in the urban environment. Bertoni and Lombardi Neto (2005) state that a decrease in porosity can increase rainwater runoff, facilitating the concentration of flows and the possible formation of erosion processes.

Pedron et al. (2004) state that erosion processes and landslides are natural phenomena intensified by anthropogenic action, which can occur, for example, due to the wrong division of land and failure to observe its suitability for use, i.e. urban soils are affected by poor planning and suffer consequences that mainly affect the population living in unsuitable areas that should often be destined for permanent protection. Another form of soil degradation in an urban environment is pollution through substances that can affect it in a predatory way. This contamination occurs through direct or indirect contact with organic pollutants (oil derivatives) or inorganic pollutants (lead, cadmium, nickel, among others) (PEDRON et al., 2004). Finally, the soil is also seen as a vector for disease transmitters, since it acts as a support for bacteria, protozoa and other transmitters. FUNASA (1992) points to ascariasis, hookworm disease, schistosomiasis, salmonellosis and other diseases as the main ones (PEDRON et al., 2004).

Thus, Craul (1985) cited by Novais et al (2000), states that urban soils are peculiar in that they have characteristics such as the modification of their structure; compaction;

bare surface; vertical and horizontal variety; changes in temperature, the specific pH of each type of soil, the action of the organisms present, infiltration and porosity; interruption of nutrient cycling; addition of anthropogenic materials and other contaminants. Santos Junior (2008) states that, according to Curcio et al. (2004), characteristics such as lower resilience and a possible increase in the potential for aquifer contamination, as well as greater susceptibility to erosion and discrepant geotechnical behavior are also features of highly anthropized soils. Pedron et al. (2004, p. 1652) conclude:

Problems such as compaction, erosion, pollution, flooding and landslides in the urban environment can be caused by inappropriate use of the soil resource, resulting from a lack of knowledge of its behavior when subjected to urban applications. Therefore, soil must be used according to its suitability for use, observing its potential and respecting its limitations and weaknesses. In all these events, the quality of life is degraded and, in many cases, the damage is irreparable or recovery is impossible. In this sense, greater investment is needed in the study and dissemination of land use and its effects on the urban environment.

2.4.1 Technogenic deposits and anthroposols

Technogenic deposits can be considered as the result of a process of occupation of the relief, taking into account anthropic action as a direct or indirect agent in shaping the landscape. This appropriation of the relief in order to incorporate technogenic features can be included, according to Ross's (1992) classification, in the sixth taxon. Fujimoto (2005, p. 78) states:

This taxon includes the smaller forms produced by current morphogenetic processes and almost always induced by human action, such as erosion furrows, technogenic dejection cones and undermining scars; or the small relief forms that develop as a result of anthropogenic interference along the slopes, such as cuts and embankments.

Peloggia (2005) states that there are three levels of human action on the environment, capable of modifying the morphology of the landscape: the modification of the relief, the alteration of geomorphological dynamics and the creation of correlative deposits similar to those of the Quaternary (technogenic deposits), based on actions called technogenesis. Technogenesis is part of a context that proposes the designation of a new geological period: "Quaternary or Technogenic", "where society begins to take certain actions in the environment, causing significant changes in its dynamics" (GODOY et al., 2005, p. 512). For Suertegaray (1997, p. 27), "this period indicates the advent of technical activity and of man as a significant force in intervening in the construction of nature".

Technogenic deposits are classified according to the category of material deposited. They can be of the orbic type (bricks, glass, concrete, asphalt, nails, plastics, various metals, etc.), orbic (detrital material with a predominance of organic waste), spodic (the result of earthworks and silting) and dredged (from dredging watercourses) (FANNING; FANNING, 1989, cited by PELOGGIA, 1998). Oliveira (1990), cited by Peloggia (1998, p. 73), states that there are three main classes of technogenic deposits: "constructed (landfills, tailings, etc.); induced (siltation, alluvium modern, etc.) and modified (siltation, alluvium modern, etc.).) and modified ("natural 'deposits' technogenically altered by effluents, fertilizers, etc.)", representing different features of different materials introduced into the environment, which often, by incorporating the natural soil resource, modify its morphology and chemical and biological characteristics.

According to Oliveira et al. (2005) the deposits of environments modified by man, often technogenic deposits, bear witness not only to anthropized environments, but also to the processes that generate and modify the current landscapes in which they are inserted. These deposits originate according to the use and management applied to the soil.

In urban areas, the establishment of residential complexes can leave its historical records in floodplains through the accumulation of leftover building materials, earthworks, garbage, clandestine sewage, sediment accumulated along watercourses, etc. Fujimoto (2002, s.n.) states that during the process of urban expansion, a notorious fact today,

[...] a large number of new materials and equipment are introduced to accompany its population growth and economic activities, linked to the urban space. Among the main changes introduced into built space, the most common are: removal of vegetation cover, construction of new landforms, increase in buildings, increase in surface runoff, surface roughness, concentrated release and accumulation of particles and gases in the atmosphere and production of artificial energy, modifying natural elements such as climate, air, vegetation, landforms and water.

In this sense, Perusi and Bergamaschi (2008), analyzing the technogenic deposits along the Furnas stream, in the municipality of Ourinhos/SP, found that 19.55 % of the materials found at strategic points in the stream are of technogenic origin, coming from the process of transporting and depositing remains of construction materials, solid urban waste, among others, configuring a picture of degradation of the alluvial plain of the stream.

Linked to this concept is Embrapa's (2004) proposal to create a new class of soils - Anthroposols. According to Pedron and Dalmolin (2002) these soils are formed due to the most diverse transformations and interventions imposed by man on the environment. Curcio et al. (2004, p. 10-11) quote:

However, man, through his profound interventions in the environment, has generated over thousands of years, on an increasing scale, volumes of soil with very different characteristics from natural soils. These volumes have very different characteristics, due to the different nature of their constituents, compositional techniques and time of formation. As a result of this wide variation, it is to be expected that they will have very different peculiarities, giving them very variable forms of use, potential and weaknesses. In addition, the vast majority of them behave differently from natural soils, above all because they have lower resilience capacities, increased potential for aquifer contamination and susceptibility to erosion, as well as differing geotechnical behavior.

These authors conceptualize Anthroposols as:

Volume formed by several or just one anthropic layer, provided it is 40 cm or more thick, made up of organic and/or inorganic material in different proportions,

formed exclusively by human intervention, overlying any pedogenetic horizon, or rock saprolite, or unweathered rock. They are volumes with very variable morphology due to the nature of their constituent materials, compositional techniques and time of formation. In general, they show a small degree of evolution, characterized by a small pedogenetic relationship between the layers. The ionic saturation of the sorptive complex is quite variable and depends mainly on the type of material used in its formation, as well as the characteristics of the remaining soil material. The presence of toxic and septic materials in its composition is very common. Drainage is very diverse and is directly related to the nature and quantity of the constituent materials, the structuring technique used to form the volume, as well as the environment in which it is deposited (CURCIO et al., 2004, p. 21).

De Kimpe et al. (2000) cited by Pedron and Dalmolin (2002) state that urban soils are generally characterized by the presence of materials of anthropic origin such as "remains of buildings, bricks, rubbers, pieces of wood, iron, concrete, remains of sidewalks and plastic materials, as well as the possible existence of pipes from the drainage, sanitary and electrical networks" (PEDRON; DALMOLIN, 2002, s.n). In addition, it is common for horizons to be removed or added to these soils, which can alter their physical and chemical aspects.

Due to these characteristics, Curcio et al. (2004) state that this is currently the most widespread class of soil in the world, and because of this proportion of occurrence, interest has grown in studying these soil formations so that it is possible to make use of them and the types of use that can be attributed to them, since they can present, as characteristics, decapitated horizons, variation in physical and chemical properties, among other discrepant characteristics. Hence the need to create this class of soil and classify them in such a way that it is possible to identify each one in a taxonomic order. The aforementioned authors report that when only one of the diagnostic conditions is found, such as inversion, removal or mixing of horizons; the presence of anthropogenic, toxic and septic materials and granulometric and chemical changes in the landscape, it is possible to imply clear signs of anthropogenesis to this soil.

It is important to note that Embrapa (1999) categorizes horizons/layers as anthropic A because they bring intense changes to their original characteristics. To this end, according to Curcio et al. (2004, p. 19-20), anthropic materials and anthropic layers are, respectively:

[...] materials of a mineral or organic nature produced by human activity, involving, among others, plastics, paper, bones, glass, ceramics, concrete, plastering materials, limestone and various types of packaging. [...] a layer less than 40 cm thick, resulting from exclusively human-induced structuring, identified in both surface and subsurface situations.

Based on this discussion, it is important to note that the concepts of anthroposols and technogenic deposits are similar. However, the origin of these two definitions goes back to different schools. Geology uses the concept of technogenic deposits which, according to Rohde (1996) cited by Rossato and Suertegaray (2000), alludes to the emergence of the Quinàrio, a period characterized by the superimposition of man on nature, that is, human activity would be such that it would go beyond the limits of geological time and imply its anthropic formations on the landscape. Agronomy uses the concept of anthroposol, which according to Embrapa (2004) refers to the modification of the physical, chemical and biological characteristics of soils due to degrading human activities.

It is clear, then, how important it is to include the soil class Anthroposols and how it is directly related to technogenic deposits, since the two classifications are similar due to the presence of degrading human action, in addition to the fact that the alterations that occur are characterized by the degeneration of the natural soil resource, transforming it into unstable environments and often unsuitable for cultivation, construction, among other uses, modifying the natural aspects of urban or rural soils. The importance of effective environmental legislation based on coherent principles is also evident, i.e. legislation capable of protecting the environment as a whole.

3 Environmental legislation

In relation to the context in which the degraded area located on Avenida Vitalina Marcusso is inserted and the constant environmental degradation that is recognized today, the importance of elements that explain, justify and play the role of protecting the environment as a whole is evident. It is in this sense that Environmental Law is used to support issues relating to this topic. Milaré (2001), quoted by Phillipi Jr. and Rodrigues (2005), states that law is a very important element when it comes to controlling a game of interests whose main objective is the appropriation of the most varied natural resources, using tools such as penalties and official impositions in an attempt to establish a minimum of order when protecting environmental goods, without neglecting the consequences that erroneous appropriation implies for both the human and natural environments. Machado (2006, p. 129130) states that this nuance of the law is responsible for:

[...] the articulation of legislation, doctrine and jurisprudence concerning the elements that make up the environment (...). It is no longer a question of building a water law, an air law, a soil law, a forest law, a fauna law or a biodiversity law. Environmental law does not ignore what is specific about each subject, but seeks to interconnect these issues with the mortar of the identity of the legal elements of prevention and reparation, information, monitoring and participation.

From then on, all social relations and human actions directed at both the intimate and the environment are, in some way, permeated by the apparatus of available laws that regulate and systematize socially constructed rules and that are interconnected to the most varied actions carried out by people. With regard to environmental issues, it is necessary for society and the state itself to recognize the viability of public authorities' applications and decisions.

In this way, man is recognized as an integral part of nature, which must seek harmony between the relationships that exist in the human and natural environments. It is in relation to the appreciation and support of nature and natural resources that the need arose to systematize a series of measures to protect them, as well as penalize individuals who disrespected the balance or improperly enjoyed the environment and its elements. This systematization reflects a step forward with regard to environmental issues, and even though many issues need to be discussed and re-evaluated, the existence of parameters governing environmental issues is extremely important.

Valuing the protection of natural resources, based on their proper use and management, comes from concern about the quality and possible scarcity of these resources, characterizing it as a new way of looking at environmental issues. For Moreira (1997), this new nuance is recent and has gained worldwide prominence since 1970, when the United Nations (UN) proclaimed it the Year of the Environment, culminating in successive conferences that discussed the future of the planet and economic development combined with environmental preservation.

Moreira (1997) stated that, since then, there have been different guiding positions on the environmental issue. These approaches offered different solutions to environmental problems:

[...] sometimes the halting of population growth, sometimes the halting of economic growth, sometimes the correction of environmental damage, sometimes the human de-occupation of some ecosystems, sometimes the redistribution of power and productive resources, sometimes environmental and social sustainability. However, these approaches have in common the same concept of the environment, i.e. the relationship between humans and nature in order to preserve natural resources (MOREIRA, 1997, p.2).

As a result of discussions on issues relating to the environment and its conservation, laws have been introduced in Brazil to ensure the validity of the right to a balanced environment, as provided for in the current Federal Constitution (FC). Article 225, Chapter VI of the aforementioned Constitution contains the legal proposals on the environment:

Everyone has the right to an ecologically balanced environment, an asset of common use to the people and essential to a healthy quality of life, imposing on the public authorities and the community the duty to defend and preserve it for present and future generations.

§ Paragraph 1 - In order to ensure the effectiveness of this right, it is the responsibility of the public authorities to: I - preserve and restore essential ecological processes and provide for the ecological management of species and ecosystems;

II - preserve the diversity and integrity of the country's genetic heritage and supervise entities dedicated to the research and manipulation of genetic material;

III - define, in all units of the Federation, territorial spaces and their components to be specially protected, with alteration and suppression being permitted only by law, with any use that compromises the integrity of the attributes that justify their protection being prohibited;

IV - require, in accordance with the law, for the installation of a work or activity that potentially causes significant environmental degradation, a prior environmental impact study, which shall be made public;

V - control the production, marketing and use of techniques, methods and substances that pose a risk to life, quality of life and the environment;

VI - promote environmental education at all levels of education and public awareness of environmental preservation;

VII - to protect fauna and flora, prohibiting, in accordance with the law, practices that jeopardize their ecological function, cause the extinction of species or subject animals to cruelty.

§ Paragraph 2 - Anyone who exploits mineral resources is obliged to restore the degraded environment, in accordance with the technical solution required by the competent public body, in accordance with the law.

§ Paragraph 3 - Conduct and activities considered harmful to the environment shall subject offenders, whether natural or legal persons, to criminal and administrative sanctions, regardless of the obligation to repair the damage caused.

§ Paragraph 4 - The Brazilian Amazon Forest, the Atlantic Rainforest, the Serra do

Mar, the Mato Grosso Pantanal and the Coastal Zone are national heritage, and their use will be made, in accordance with the law, within conditions that ensure the preservation of the environment, including the use of natural resources.

§ Paragraph 5 - The lands that are vacant or collected by the states through discriminatory actions and that are necessary for the protection of natural ecosystems are unavailable.

§ Paragraph 6 - Plants operating with a nuclear reactor must have their location defined by federal law, failing which they cannot be installed.

This law makes clear the responsibilities of the public authorities when it comes to maintaining the quality of life of the population, ensuring the basic conditions for caring for the environment, as defined by the Federal Constitution. In relation to this law, it can be said that its text contains constitutional duties, which most of the time are not respected either by the public authorities themselves or by the society involved. An example of this is the increasing degradation of natural resources, expressed in the form of the death of springs, the establishment of erosion, silted watercourses, soil infertility, land displacements, among others. Because of the duties contained in this law, it was necessary to set up collegiate bodies, whose competencies are to help formulate guidelines for government action, as well as to deepen the discussion about rights and duties towards the environment and the maintenance of its quality. Thus, the legislation instituted by the Forest Code, Law No. 4771 of September 15, 1965, in its articles 2 and 3, points to the following as permanent preservation areas (APP):

Art. 2 - For the sole purpose of this Law, forests and other forms of natural vegetation located in:

a) along rivers or any other watercourse, in a marginal strip whose minimum width shall be:

1 - 5 (five) meters for rivers less than 10 (ten) meters wide: 2 - equal to half the width of rivers measuring from 10 (ten) to 200 (two hundred) meters between their banks;

3 - 100 (one hundred) meters for all courses with a width of more than 200 (two hundred) meters.

b) around ponds, lakes or natural or artificial water reservoirs;

c) in springs, even in the so-called "eyes of water", whatever their topographical location;

Figure 2: Urban erosion "recovery" work on Avenida Vitalina Marcusso, in the municipality of Ourinhos/SP - a waste of public money and environmental degradation

d) on the tops of hills, mountains and mountain ranges;

e) on slopes or parts thereof with a gradient of more than 45°, equivalent to 100% on the line of greatest gradient;

f) in the restingas, as dune fixers or mangrove stabilizers;

g) on the edges of the trays or plateaus;

h) at an altitude of more than 1,800 (one thousand eight hundred) meters, in natural or artificial fields, native forests and countryside vegetation.

Art. 3 - Forests and other forms of natural vegetation designed to: a) mitigate land erosion shall also be considered permanent preservation, when so declared by act of the Public Authorities;

b) fixing the dunes;

c) to form buffer strips along roads and railroads;

d) to assist in the defense of national territory at the discretion of the military authorities;

e) to protect sites of exceptional beauty or of scientific or historical value;

f) to house specimens of fauna or flora threatened with extinction;

g) to maintain the environment necessary for the life of forestry populations;

h) to ensure conditions of public welfare.

According to this law, springs and/or waterholes, as well as riparian vegetation, are considered APPs. In criticism of the Forest Code, there is a need to improve the legislation with regard to urban areas. It is understood that in the articles presented in the Code, there is no distinction between APPs and their limits in urban and rural areas. However, due to the intense waterproofing process that cities are undergoing, the importance of maintaining APPs is clear. We also consider the minimum limits established by this code, since in urban areas the dimensions presented encounter various barriers when they are implemented, and often all the riparian vegetation is eliminated.

Araùjo (2002, p. 3) states that Permanent Preservation Areas (PPAs) must be protected, and are characterized by the imposition of the law regarding the

[...] preservation of water resources, geological stability and biodiversity, as well as the well-being of human populations The protection regime for APPs is quite strict: the rule is untouchability, with the exception of the suppression of vegetation only in cases of public utility or social interest that are legally provided for.

For the establishment of these areas, the legislation provided for is Federal Law No. 4.771 of 1965, the Forest Code, which recognizes these areas as:

II - permanent preservation area: an area protected under the terms of articles 2 and 3 of this Law, whether or not covered by native vegetation, with the environmental function of preserving water resources, the landscape, geological stability, biodiversity, the gene flow of fauna and flora, protecting the soil and ensuring the well-being of human populations; (Included by Provisional Measure No. 2.166-67, of 2001).

In this legislation, two categories of APPs can be recognized, cited by Araùjo (2002) as being those created by the law itself and those that require an exclusive declaratory act by the Public Power for their determination.

The law also includes a Sole Paragraph, added to the Forest Code at Federal Law No. 7083 of 1989 (BRASIL, 2009), explaining the need for joint action by the public authorities to create PPAs and the capacity of this segment to create PPAs.

In the case of urban areas, understood as those included in the urban perimeters defined by municipal law, and in metropolitan regions and urban agglomerations, throughout the territory covered, the provisions of the respective master plans and land use laws shall be observed, respecting the principles and limits referred to in this article.

Article 3 of the Forest Code complements the above observation.

Art. 3. Forests and other forms of natural vegetation designed to: a) mitigate land erosion shall also be considered permanent preservation, when so declared by act of the Public Authorities;

b) fixing the dunes;

c) to form buffer strips along roads and railroads;

d) to assist in the defense of national territory at the discretion of the military authorities;

e) to protect sites of exceptional beauty or of scientific or historical value;

f) to house specimens of fauna or flora threatened with extinction;

g) to maintain the environment necessary for the life of forestry populations;

h) to ensure conditions of public welfare.

An example of environmental councils is the National Environmental Council (CONAMA), an advisory and deliberative body governed by the National Environmental System (SISNAMA) and created at the same time as Law 6.938/81, which sets out the National Environmental Policy.

According to the Ministry of the Environment (2010), CONAMA is responsible for monitoring compliance with the rules, imposing fines and penalties for actions that damage the environment, rules and standards relating to noise pollution and natural resources, encouraging closer ties between the other collegiate environmental bodies, among other measures. Among the norms and criteria established by CONAMA, in conjunction with IBAMA and other collegiate environmental bodies, is environmental licensing, which must be preceded by a Preliminary Environmental Impact Study (EPIA or EIA).

These legal tools are intended for preventive application, so that the effect of human interventions on the physical environment can be assessed beforehand. In this way, there are legal resources to support environmental management, making technical and administrative points of view feasible. These actions are part of an environmental planning policy, as well as the search for harmony between human and natural communities (SOUZA, 2003). Even so, predatory actions against the environment culminate in a wide range of environmental impacts. According to Article 1 of Resolution 01/1986 - CONAMA, environmental impact is nothing more than:

[...] any alteration of the physical, chemical and biological properties of the environment, caused by any form of matter or energy resulting from human activities that directly or indirectly affect it:

I - the health, safety and well-being of the population; II - social and economic activities;

III - the biota;

IV - the aesthetic and health conditions of the environment; V - the quality of environmental resources.

Capelli (1994, s/p.) refers to environmental impact as:

The set of consequences of the creation or presence of an enterprise on the environment or the set of repercussions and consequences that a new activity or new work, whether public or private, may cause to the physical environment with all its components (security of the territory) and to the living conditions of the population concerned (quality of life).

This means that environmental impact means not only pollution of the environment, but also a wide variety of scars on the landscape. Activities that involve drastic alterations to the environment are considered to generate environmental impact, such as construction work that requires a large amount of earth movement, landfills, mineral extraction, gullies and gullies, deforestation, among other actions and representations whose result would be detrimental to the natural balance and its maintenance. The

Preliminary Environmental Impact Study (EPIA or EIA) seeks to prevent the negative responses that a project can cause on the earth's surface, and is considered an essential tool for environmental, urban and industrial planning. This instrument had its origins in the United States of America, through a decree-law in 1969, which validated the National Environmental Policy Act (NEPA), the first legal document to discuss the aspects that the EIA would cover.

From the 1970s onwards, this model spread to other countries around the world, with the aim of guaranteeing environmental protection, including the participation of society in discussions and decisions to be made regarding the implementation of activities that could harm the environment (MOREIRA, 1997; GOULART; CALLISTO, 2003). In Brazil, the EIA was introduced by Law No. 6938/81, which sets out the National Environmental Policy. However, there was a gap, since the need for an EIA prior to the start of impacting works or activities was not expressed. CONAMA Resolution 001/86 established this obligation for preventive action. Thus, according to Benjamim (1993, p. 15):

No other environmental law institute better exemplifies this preventive approach than the EIA. It was precisely in order to foresee (and thus prevent) damage before it manifested itself that the EIA was created. Hence the need for the EIA to be drawn up at the right time: before the start of the project's execution, or even preparatory acts. No wonder the Federal Constitution chose to rename the institute from "environmental impact assessment" to "prior environmental impact study".

Thus, according to Souza (2003, p 25), the objective of the EIA or EPIA is:

[...] evaluate the proportions of the possible alterations that an undertaking, whether public or private, may cause to the environment, in order to effectively guarantee the right enshrined in Article 225 of the Federal Constitution, 'the right to an ecologically balanced environment'. (...) The EIA, in short, is nothing more than a study of the probable changes in the various socio-economic and biophysical characteristics of the environment that may result from a proposed project.

Souza (2003, p.31) also affirms the importance of the EIA in the sense of:

Qualifying and, where possible, quantifying the environmental impact in advance is the role reserved for the EIA as a support for proper planning of works or activities related to the environment. It is true that predicting the harmful effects of a project can often be very tricky, as some changes to the ecological balance only appear very late. Hence the correct consideration of the EIA as an administrative procedure for preventing and monitoring environmental damage.

Article 2 of Resolution 001/86, instituted by CONAMA, as well as the amendments added to this article by Resolutions 011/86 and 05/87 of the same body, define the works and activities that need to undergo a prior study. This law states:

An environmental impact study and respective environmental impact report (RIMA) must be prepared and submitted for approval to the competent state agency and IBAMA, on a supplementary basis, for the licensing of activities that modify the environment, such as:

I - Roads with two or more lanes; II - Railways;

III - Ports and terminals for ore, oil and chemical products;

IV - Airports, as defined in item 1, article 48, of Decree-Law No. 32, of 18.11.66;

V - Oil pipelines, gas pipelines, mining pipelines, sewage collection trunks and outfalls;

VI - Electricity transmission lines above 230KV;

VII - Hydraulic works for the exploitation of water resources, such as: dams for hydroelectric purposes, above 10MW, sanitation or irrigation, opening of canals for navigation, drainage and irrigation, straightening of watercourses, opening of bars and inlets, transposition of basins, dykes;

VIII - Extraction of fossil fuels (oil, shale, coal);

IX - Ore extraction, including class II, as defined in the Mining Code;

X - Landfills, processing and final disposal of toxic or hazardous waste;

XI - Electricity generation plants, regardless of the primary energy source, above 10MW;

XII - Industrial and agro-industrial complexes and units (petrochemicals, steel, chlorochemicals, alcohol distilleries, coal, extraction and cultivation of water resources);

XIII - Industrial districts and strictly industrial zones - ZEi;

XIV - Economic exploitation of timber or firewood, in areas over 100 hectares or smaller, when it reaches significant areas in percentage terms or in terms of environmental importance;

XV - Urban development projects over 100 hectares or in areas considered to be of relevant environmental interest at the discretion of SEMA and the competent municipal and state bodies;

XVI- Any activity that uses charcoal, derivatives or similar products, in quantities exceeding ten tons per day.

XVII - Agricultural projects covering areas of more than 1,000 ha. Or smaller, in this case, when dealing with significant areas in percentage terms or of importance from an environmental point of view, including areas of environmental protection.

XVIII - Projects potentially harmful to the National Speleological Heritage;

With regard to the EIA/RIMA, this must be drawn up, in accordance with article 7 of CONAMA Resolution 001/86, by a multidisciplinary team capable of developing diagnoses and analyses of the proposed project, and the results obtained are the sole responsibility of the technicians involved. To this end, according to Souza (2003), a diagnostic document should be drawn up describing the situation of the site before the project is implemented, covering considerations of both physical aspects, such as water resources, existing fauna and flora, climatic conditions, interactions between ecosystems, among others; and socio-economic aspects, highlighting relationships and dynamics such as land use, population, health, education, among others. An analysis must also be created that takes into account issues related to the positive and negative impacts that the project may cause, and permission to carry out the implementation works or activities, environmental licensing, is only granted after the EIA has been presented and thoroughly analyzed, taking into account the aspects raised and the solutions proposed for possible damage to the environment, damage that may even be irreversible.

As an essential part of the Preliminary Environmental Impact Assessment, contained in Article 9 of CONAMA Resolution 001/86, is the Environmental Impact Report: a note that contains the considerations of the technicians involved in producing the EIA. This document must contain visual aids such as graphs, maps, among others, which make it easier to interpret the environmental consequences and the alternatives indicated, as well as the (in)feasibility of each one. This report must be disseminated to society as a whole, based on the Principle of Information in Environmental Law. Mirra (1999) states that this disclosure through public hearings is essential if the objectives of the EIA/RIMA and environmental licensing itself are to be achieved, since the participation of the society involved in all phases of the study's development must be considered imperative.

From there, it is possible to present and discuss environmental licensing. Article 170, VII, sole paragraph, of the Federal Constitution guarantees everyone "the free exercise of any economic activity, independent of authorization from public bodies, except in the cases provided for by law". This means that certain activities or works require permission to be carried out, and this will only be possible if provided for by law: this is the case with undertakings and activities considered to cause actual or potential environmental degradation of some kind. In this context, according to Machado (2006), environmental licensing is an authorization created by law, given by a competent body and which needs to be renewed so that a certain licensed activity can continue to exist. It is clear that the Public Administration can intervene from time to time to check the environmental quality of licensed activities. It is important to point out that the authorization or renewal of environmental licensing must be carried out in accordance with parameters that take into account standards and guidelines that avoid as much environmental damage as possible, as is the case with pollutant emission standards. In this way, the Public Administration's analysis of the licensing application must respect these guidelines, always aiming for the minimum impact on the environment. It is also pointed out that, according to Antunes (1996), if it is necessary to modify the environmental standards of a given activity, this can only be done after the license has expired, and that if the authorization granted requires any requirement and this is not complied with, the license issued may lapse, as there has been negligence on the part of the entrepreneur with regard to concern for preventing environmental damage and maintaining the quality of the environment.

It is then defined in CONAMA Resolution 273/97 that environmental licensing and environmental license are, respectively:

Administrative procedure by which the competent environmental body licenses the location, installation, expansion and operation of undertakings and activities that use environmental resources, considered to be effectively or potentially polluting or those that, in any form, may cause environmental degradation, taking into account the legal and regulatory provisions and technical standards applicable to the case.

An administrative act by which the competent environmental agency establishes the conditions, restrictions and environmental control measures that must be complied with by the entrepreneur, whether an individual or a legal entity, in order to locate, install, expand and operate undertakings or activities that use environmental resources that are considered to be effectively or potentially polluting or those that, in any form, may cause environmental degradation.

The same resolution provides for the issuing of three types of licenses, respecting the order in which the project is installed. The Preliminary License (LP) corresponds to the planning phase, containing the basic conditions for the location, installation and operation processes. The Installation License (LI) authorizes the installation process upon approval of a project containing the project's specifications. After analysis and approval of an operating plan, the Operating License (LO) will be issued; it will authorize the start of the licensed activity and the use of pollution control equipment. These licenses are granted gradually, according to the evolution and appropriate conditions for the maintenance, allocation and operation of the project.

CONAMA Resolution 237/97 also lists a series of activities that require the issuance of an environmental license, provided that they cover undertakings involving: extraction and treatment of minerals; non-metallic mineral products, metallurgical, chemical, mechanical, wood, rubber, hides and skins, food and beverage, tobacco industries; concrete, asphalt and electroplating plants, among others; construction projects such as highways, railroads, waterways, dams and dykes, among others; services such as thermoelectric and hydroelectric power generation and transmission, water treatment plants, among others.

From then on, it is clear that not only activities that would have a major impact on the environment must obtain environmental licensing from the competent body, but also all those that could potentially degrade or pollute the environment. This fact, according to Prado Filho and Souza (2004), citing Van Acker (1998), demonstrates a profound change in the guidelines and positioning of the function of environmental licensing, since in addition to covering a series of activities that potentially degrade the environment, it also establishes the EIA/RIMA as a requirement for such undertakings, and the studies carried out will be determined by the responsible environmental agency with the participation of the entrepreneur. The Preliminary Environmental Impact Study is therefore an essential tool for Environmental Licensing. This obligation reinforces and validates environmental concerns in Brazil, and although success has been achieved in several cases, it must be understood that the effectiveness of these processes depends on the action and participation of qualified professionals who, together with the public authorities and civil society, play roles aimed at defending the environment, articulating a set of actions that prevent corruption and the approval of works and activities that are harmful to the environment. In this way, we can see the unquestionable importance of studies on this subject, as well as the support of the relevant legislation, since there is still a need for implementation to make it fully effective.

4 Planning as a technique for preventing environmental damage

One of the measures that public authorities are responsible for in terms of preventive action when it comes to the degradation of natural resources and the environment as a whole is the technique of planning. One of the characteristics of planning is that it should be based on decisions that take into account the living conditions of present and future populations, making it possible to preserve the environment. According to Silva (2003, s/p):

[...] planning must make decisions that optimize the future. Planning requires integrated studies that select and systematize information about the area or place where it will be applied and identify the state of the environment, the pressure exerted on it and the response given by society to improve it. The need for development to go hand in hand with actions that conserve nature brings with it the need to look for strategies capable of collecting, weighing up and integrating data on the components of ecosystems.

From this point on, the importance of planning becomes clear, and this is made possible through studies and tools that materialize its foundations. In order to carry out environmental planning of unquestionable quality, it is necessary to have sound professional training that allows competent professionals, together with public or private organizations, to encourage research that combines environmental, economic, social, cultural and political issues in an integrated manner, so as to allow studies that complement and assist in decision-making by the public authorities.

For Ferreira (1999, p. 1582), planning means:

The act or effect of planning; the work of preparing any undertaking, according to determined routes and methods, planning; the process that leads to the establishment of a coordinated set of actions (by the government, the management of a company, etc.) aimed at achieving certain objectives; drawing up government plans or programs, especially in the economic and social field.

From this point on, planning is understood as preparation for decision-making. It must be continuous and present a multidisciplinary approach, integrating the most varied spheres of analysis, as well as offering alternatives that make it possible to make the right choice of actions that culminate in an effective result. Van Lier (1994) cited by Silva (2003) presents two dimensions of planning: the first concerns land use policies, similar to zoning; the second deals with improvements in the spatial and physical aspects of the environment, generally dictated by the first dimension. For Almeida (1993), the most important dimensions of study in planning refer to the analysis of the socio-economic issues of a given population, the so-called demand, since it is through this that thc objectives to be achieved are outlined. Studies that define the characteristics of the physical environment are also very important, as it is from these that human activity develops. The two authors share similarities in what they say, pointing to studies that integrate physical and environmental aspects with the political, socio-economic and cultural conditions of a given society.

According to Floriano (2004, p. 12), planning is a broad management tool that takes the following aspects into account:

It is a process of organizing tasks to achieve a goal, with characteristic and sequential phases that are generally in the following order: identifying the object of planning, creating a vision of the subject, defining the planning objective, determining a mission or commitment to achieve the planning objective, defining work policies and criteria, setting goals, developing a plan of actions necessary to achieve the goals and fulfill the mission and objectives, establishing a system for monitoring, controlling and analyzing the planned actions, defining a system for evaluating the controlled data and, finally, providing for measures to be taken to prevent and correct any deviations that may occur in relation to the plan.

As far as environmental planning is concerned, according to Christofoletti (1999), geographic knowledge must be applied to it, as there is a need to highlight ecosystems and geosystems, seeking different perspectives on environmental systems when carrying out works and undertakings. It is clear that environmental planning must present methodologies that consider the actions and consequences of a given site's capacity for use and the actual uses applied to it, always aiming for the least impact and degradation of the environment through a systemic and integrated approach.

One of the techniques used in the management and planning system is zoning. According to Machado (2006, p. 187), zoning consists of:

It divides the territory into parcels in which certain activities are authorized or the exercise of other activities is prohibited, either absolutely or relatively. Although zoning alone is not the solution to all environmental problems, it is a significant step forward.

Milaré et al. (2004) point out that the objectives of zoning should be in line with the National Environmental Policy, promoting sustainable development based not only on political and administrative processes, but also by combining technical and scientific knowledge in order to stimulate the conscious use of natural resources and territorial planning in favor of environmental quality.

The technique of zoning is applied through criteria in order to divide a given region into zones, and the result of the work can be presented in the form of thematic maps. There are various types of zoning, some of which are designated by law. This is the case, for example, with urban zoning, ecological-economic zones and environmental zoning. It can be said that the above categories can be implemented together, since one complements the other. When inserted into an urban context, they bring up various discussions, presenting specific dimensions for each defined zone.

When it comes to ecological-economic zoning, its main application is in Environmental Protection Areas (APAs). Silva (2003) mentions that in order for this type of zoning to achieve its objectives, work must be carried out in accordance with the most varied local conditions, above all valuing the ecological balance of the area. According to Programa (2001, p. 17) quoted by Silva (2003, p.24), "the EEZ has been, in recent years, the Brazilian government's proposal to support social, economic and environmental planning decisions for the development and use of the national territory on a sustainable basis".

Environmental zoning, according to Montano (2002), should be seen not only as a process of parceling out a given area by allowing or prohibiting certain uses and occupations. For this author, who bases his statement on the words of Macedo (1998), the preparation of environmental zoning must be based on a development policy that aims to improve and maintain a given territory, i.e. if used correctly, this tool is capable of promoting various positive changes to the management system and to society as a whole. Floriano (2004) also points to environmental zoning as part of urban and rural physical planning, so that it brings more than just a plan for land use and occupation to the two sides, pointing out solutions and respecting the potential of each location.

As far as urban zoning is concerned, it opens up discussion about the division of land in a city through laws on land use and occupation, mainly instituted in the Master Plan and the Statute of Cities. Silva (2003, p. 21) states that:

Urban zoning, which is always defined in the urban land use and occupation legislation of each municipality, regulates the forms of use and types of construction in the urban area, determining what can and cannot be done in each zone. Generally, urban areas are divided into residential zones, commercial zones, industrial zones, central zones, protection zones, expansion zones and special zones. This zoning should support the city's master plan, where the main urban planning guidelines are defined.

Carvalho and Braga (2001) state that urban zoning is the most widespread and most criticized urban planning instrument, since it has undesirable indirect responses such as real estate speculation and socio-spatial segregation, since it defines the use and occupation of urban land, segregating the types of use. In this way, it also segregates the social groups that carry out activities in each of these determined uses.

These three types of zoning together support and regulate activities in a municipality. Their application seeks to mitigate erroneous forms of occupation and construction on urban land, with a view to the well-being of the population and the balance of ecosystems. When used appropriately, backed up by other legislation concerning the environment and its protection, its effectiveness is evident.

However, it is important to emphasize the need for further work on this subject, since the proper application and use of these tools also comes from improving techniques and their widespread information. Zacharias (2006, s/n) adds:

Zoning defines spaces according to pre-established grouping criteria, which usually express the potential, vocations, restrictions, weaknesses, susceptibilities, successes and conflicts of a territory. Planning, on the other hand, establishes guidelines and goals to be achieved within a timeframe, relating to these thematically delineated and represented spaces.

4.1 The Master Plan and its role in urban development and expansion

All the environmental impact prevention and control techniques presented here are intended to bring reality and legislative regulation to the municipal level, such as the Organic Laws and the Master Plan of a municipality. With the approval of the City Statute, federal law no. 10.257/2001, which regulates in the sole paragraph of its article 1 the "public order and social interest rules that regulate the use of urban property in favor of the collective good, safety and well-being of citizens, as well as environmental balance", the bases for the implementation of the Master Plan were defined, According to item I of article 42, it must contain at least "the delimitation of the urban areas where compulsory parceling, building or use may be applied, taking into account the existence of infrastructure and demand for the use of art. 5 of this Law". This article deals with the specifics of "compulsory subdivision, building or use of unbuilt, underutilized or unused urban land, and must establish the conditions and deadlines for implementing this obligation".

As for the obligation, this is spelled out in article 41 of Law 10.257/2001, items I, II, III, IV and V. The cities whose Plan is compulsory are:

I - with more than 20,000 (twenty thousand) inhabitants;

II - members of metropolitan regions and urban agglomerations;

III - where the Municipal Government intends to use the instruments provided for in § 4 of art. 182 of the Federal Constitution;

IV - areas of special tourist interest;

V - within the area of influence of projects or activities with a significant regional or national environmental impact.

However, the state of Sao Paulo has extended this obligation to all its municipalities. Carvalho and Braga (2001) are critical of this, as it runs counter to municipal autonomy. However, despite all the controversy arising from this issue, the state's municipalities, including those with a population of less than twenty thousand inhabitants, have made the Master Plan compulsory in their Municipal Organic Laws.

Other important points raised by the City Statute and which form part of the content and principles of the Master Plan, are the definition of guidelines for the delimitation of urban areas that can be implemented through the Concessionary Right to Build; areas subject to the application of consortium urban operations; authorization for the transfer of the right to build by urban property owners; and the existence of a system for monitoring and controlling the execution of the plan. The master plan must encompass the rural and urban areas of a municipality, and the participation of society or its representatives in its formulation is extremely important; this is a right guaranteed by the City Statute (paragraph 4, art. 40, inc. I) and it is through this participation that urban planning is democratized. Another important factor is that the Plan is a management tool that must ensure the transparency of urban policy, publicizing the guidelines of this policy so that society has access to it and can evaluate its proposals and foundations. Carvalho and Braga (2001, p. 98) add:

The Master Plan is an eminently political instrument, whose primary objective should be to provide transparency and democratize urban policy, i.e. the Master Plan should be, first and foremost, an instrument of democratic city management. In this sense, it is important to emphasize these two aspects of the Plan: transparency and democratic participation.

Thus, this document becomes the most important instrument when it comes to urban development and expansion policy, and it is the City Statute that regulates its principles and contents. Its functions, then, are to optimize the planning of urban space by regulating the use and occupation of land, with a view to improving living conditions for the population and protecting the environment, as well as instituting democratic planning and management in the municipality through the active participation of the society involved.

4.2 PAC projects and the Master Plan for the municipality of Ourinhos/SP: considerations regarding environmental protection policy

According to DIEESE (2007) cited by Ricardo et al. (s/a), the PAC is nothing more than a set of measures that are articulated through goals set by the government to boost the country's economic growth between 2007 and 2010, through investments organized in three areas, which according to Ricardo et al.

Ricardo et al. (s/a, p. 7) are:

Logistics involves the construction and expansion of highways, railroads, ports, airports and waterways; Energy corresponds to the generation and transmission of electricity, production, exploration and transportation of oil, natural gas and renewable fuels; the Social and Urban Area includes sanitation, housing, subways, urban trains, universalization of the Light for All program and water resources. The investments planned for Logistics amount to R$ 58.3 billion; Energy, R$ 274.8 billion and for Social and Urban, R$ 170.8 billion, totaling R$ 503.9 billion in public and private investments.

Kasznar (2008, p. 4) divides the measures for economic growth and infrastructure into five blocks of action. These aim to project the country towards a return to development, and are used as a banner by the current government. These blocks are

The first block refers to investments in infrastructure, especially for energy production, especially oil, where the government is investing more of its resources. The second block is to stimulate credit and financing. The third is about improving the regulatory framework and environmental laws. The fourth block aims to reduce and improve taxes. The fifth block proposes long-term fiscal measures to give macroeconomic consistency and sustainability to the proposed set of measures.

In this way, the PAC is an attempt to stimulate Brazil's economic growth, sustaining it on modern technological bases so that it can be effectively integrated into the international market, as well as reducing the levels of social inequality and improving the living conditions of the country's underprivileged population, since it covers areas such as basic sanitation and transportation. The need for these proposals is unquestionable, however, it is important to question the budget for these projects and the seriousness with which they are treated, given that large sums of public money and private investment are being invested.

The municipality of Ourinhos is one of the 58 municipalities that have been included in the PAC package of works. 55.9 million reais will be invested in urban infrastructure improvements, 50.3 million reais of which will be financed by the federal government through the Sanitation for All program, for micro-drainage and sanitation works. According to the municipality's urban planning secretary, the projects to be subsidized by PAC have been under preparation since 2005. According to the Ministry of Cities (2008), the start of the PAC project in Ourinhos was authorized on May 6, 2008, benefiting more than 28,000 families living in the municipality, and around 40 million reais in investments are budgeted for 2010 (PREFEITURA MUNICIPAL DE OURINHOS, 2010).

The Master Plan for the municipality of Ourinhos was drawn up during the administration of Mayor Toshio Misato and instituted by Complementary Law No. 499 of December 28, 2006. This document is responsible for establishing the guidelines for action by public and private agents with regard to urban planning and management policies and guarantees socio-economic development in line with environmental preservation. It also ensures the participation of society in the most varied processes of urban management, using the democratic participation provided for in the City Statute and the Federal Constitution. Article 4 outlines the municipality's social functions:

[...] correspond to the socially just and ecologically balanced use of the municipality's territory and the guarantee of citizens' rights to housing, environmental sanitation, infrastructure and public services, health, education, urban mobility and accessibility, work, culture, leisure, preservation of environmental and cultural heritage and the development of trade and production, with a view to socio-economic inclusion.

With regard to the environment, this law provides for the right to an ecologically balanced environment for current and future generations, through the joint work of public authorities and civil society. As for the actions of the public authorities, these are set out in Article 9, sole paragraph:

I. create a specific administrative structure to manage the Municipal Environmental Policy;

II. establish management and control mechanisms, linking up with municipal departments, especially education, transport, health, tourism and sport;

III. to train and qualify public authorities for an integrated administration that incorporates inter-sectoral dialogue between municipal departments and the productive sector;

IV. promote environmental education at all levels of education and public awareness for the preservation of the environment, in accordance with art. 225, item Vi of the Federal Constitution of 1988;

V. promoting research and technological development geared towards the rational use of natural resources;

VI. encourage public participation in the definition and monitoring of municipal development and environmental management plans for:

a. reduce pollution of watercourses;

b. prevent flooding;

c. reduce pollution and soil degradation;

d. combating noise pollution in urban and rural areas;

e. protect watercourses, springs, Permanent Protection Areas (APP's) and riparian forests, in accordance with the Forest Code;

f. protect green areas, public squares and other public facilities with large vegetation;

g. protecting the natural and scenic heritage of local interest;

VII. incorporating civil society into actions to control and enhance the municipality's environment, particularly private initiative in projects of common interest;

VIII. establish the Municipal Council for the Defense of the Environment - CONDEMA, in accordance with the guidelines of the Organic Law of the Municipality, art. 321, an advisory and deliberative body, in addition to its competencies established in its creation law, on environmental issues.

As far as this paragraph is concerned, it is possible to draw the following comparisons with the PAC projects: as far as urban streams are concerned, there were canalization projects for the Monjolinho, Jacu, Jacuzinho, Furnas and Furninhas streams, as well as paving works and the construction of rainwater galleries in various parts of the city. Among the sites that benefited from the rainwater culvert work was the degraded area located on Avenida Vitalina Marcusso. It is a private area that has undergone a process of "recovery" with public money distributed through PAC. In an interview with Gielff (2010), the architect responsible for the work in the municipality of Ourinhos, when asked about the viability of investing in private land using public money, the interviewee said that in order for the project to be approved by the competent bodies, it was necessary to make adjustments to the conduct of the public prosecutor's office. The justification for this intervention was the construction of the drainage network, without adequate sizing, which directed all the rainwater from the adjacent neighborhoods onto the property in question, resulting in the formation of urban erosion and all the installed degradation. Work on the site began in July 2009, and was paralyzed from August to November. Gielff (2009), when asked about the interruption of the work, said that it was necessary because, due to a change in the project budget by the construction company responsible (Penascal Engenharia e Construçao Ltda.), it had to go through the bureaucracy and approval of the Urban Development Support Branch Representation (REDUR) again. Because of this, and the heavy rains that occurred during this period, the galleries were completely destroyed, compromising the quality of the work and demonstrating a total disregard for public money.

The budget for this work was R$738,820.80. The concern about the use of public money is debatable, given that the conditions identified at the site show total inoperability, both on the part of the construction company responsible and of the public authorities. The new rainwater galleries would flow into the bed of the Agua da Veada stream, which would relocate the problem. Gielff (2009) stated that the other improvements, as well as maintaining the stability of the land, would be the responsibility of the owner, with the city council only responsible for channeling the rainwater. In another analysis, citing the municipal Master Plan, item VI of art. 9, "protect watercourses, springs, Permanent Protection Areas - APP's and riparian forests, in accordance with the Forest Code". It can be seen that this law does not apply to the Agua da Veada stream and its surroundings, since it is completely devoid of riparian vegetation, and the stream's springs have been completely buried due to the construction of rainwater galleries.

Taking the discussion to the context of the degraded area of Avenida Vitalina Marcusso, it can be seen that point e of item VI of the city's Master Plan provides for the same dimensions as the Forest Code. It is therefore understood that the concept of APPs would be the same as that adopted by federal legislation. In this case, there is a breach of the law, indicated by the suppression of riparian vegetation. According to Araùjo (2002, p. 3), permanent protection areas are:

[...] areas in which, by law, the vegetation must be kept intact, with a view to guaranteeing the preservation of water resources, geological stability and biodiversity, as well as the well-being of human populations. The protection regime for PPAs is quite strict: the rule is untouchability, with the exception of the suppression of vegetation only in cases of public utility or social interest that are legally provided for. In practice, however, these and other PPAs have simply been ignored in most of our urban centers, a reality that is associated with serious environmental damage, such as the silting up of water bodies, and events that pose serious risks to human populations, such as floods and landslides. It is said that the rules regulating APP are among the most poorly worked interfaces between federal environmental legislation and urban issues. The flaws in the legislation are pointed out as one of the main factors contributing to non-compliance with these rules in urban areas.

Another extremely important point when discussing the PAC projects is the need for the EIA/RIMA and Environmental Licensing to release the investment. When I put this question to the architect responsible for the PAC works in the municipality of Ourinhos, Gielff (2009) said that the EIA/RIMA was carried out by technicians from the town hall, and that the license was issued by CETESB. As explained, every RIMA must be publicized through public hearings and made available to the population. When questioned about this document, the architect said that it was of good quality and that the work had begun after approval and issuance of the license, making it clear, according to the interviewee, that all stages of the work were legally supported.

In order to give validity to his words, the Urban Development Department has made three maps available: one shows the areas of the city covered by the PAC works; the other shows the works plan for the urban rainwater drainage system for Avenida Vitalina Marcusso. On the map showing the areas benefiting from the PAC, it can be seen that all the work is on urban infrastructure: paving, sidewalks, rainwater galleries and canalization of streams. On the maps that represent the construction plan, longitudinal profiles of the terrain are shown, pointing out its level quotas and the layout of the dissipation boxes and pipes that have been built. This shows the construction project carried out by the city council and shows that the location of the new launch point for the drainage network reached the Agua da Veada stream.

The PAC works in the city of Ourinhos have taken on great dimensions and have led to discussions about their viability, especially with regard to the channeling of the streams. According to information obtained in an interview with the architect, the channeling was done using a technology called "bolsacreto".

An example of this intervention is the Monjolinho stream. Totally silted up, part of its length runs through the "Bióloga Tânia Mara Netto Silva" Municipal Ecological Park. In order to complete the work, several trees had to be removed from the park to bring in equipment and materials. At this location, the canalization will take place using a block system, which, according to the interviewee, will cause "less damage to the environment and the park". He said that the cutting down of the trees was permitted and the license issued by CETESB, and that there are projects to replace the plants removed so that there is no damage to the natural community there. However, the canalization is causing a great deal of damage to aquatic life, and could even exterminate it. This intervention, as well as the one taking place in the Furninhas stream, located in the Parque Minas Gerais neighborhood, brought several protests from the Ourinhense community, and according to Gielff (2009) the protests did not present a risk to the projects, as they already had approval. In addition, the city council intended to replant the native forest along the canalized streams, in an attempt to restore a natural environment. For all the interventions mentioned, the main issue is the responsibility of carrying out the works, since they are financed with public money, as well as involving environmental aspects.

4.3 Sustainable development

After the presentation of legislation based on the protection of natural and environmental resources and techniques and tools for preserving them, we see the need to discuss sustainable development, a controversial concept that permeates the most diverse conferences today. It's a term that has left academic discussions and is now part of people's daily lives.

According to Almeida (1993), this term has its origins in the ideas presented in the Brundtland Report, known in Brazil as "Our Common Future", a preparatory text for the United Nations Conference on the Environment held in Rio de Janeiro (Eco-92). These ideas permeate the directive to promote development while guaranteeing the needs of future generations. Carvalho (1991) points out that the term "sustainable" refers to what is in balance, and when this balance is applied to economic and social development, among others, it creates the false expectation of the possibility of the existence of a sustainable society, that is, in balance with nature and with itself.

This reflection raises questions such as: whose sustainability? Sustainability for whom? Sustainability provided and determined by whom? Who would be the agents and actors of this sustainability?

These questions start from the assumption, discussed by Almeida (1993), of the idea of sustainable development as something holistic, systemic and integrating the exploitation of natural resources, economic and technological development and society, based on parameters defined by the dominant social classes. However, Almeida (1993, p. 43) adds that:

[,,,] the discussion on sustainable development today is polarized between two main conceptions: on the one hand, the concept/idea as being generated within the sphere of the economy, with this reference being used to think about the social. In this way, nature is incorporated into the production chain (nature becomes a capital asset); on the other hand, an idea that tries to break with the hegemony of the economic discourse and the disproportionate expansion of the economic sphere, going beyond the instrumental, restricted vision that the economy imposes on the idea/concept.

With regard to these two concepts, Drizek (1992) states that the economic-liberal approach is based on self-regulation of the market, in order to combine prosperity, competition and economic growth. This combination would lead to the rational use of natural resources, requiring new techniques of exploitation, reuse, in other words, techniques that would be developed to be compatible with the demands of the environment and the market. A relevant point is that this economic-liberal approach dismisses the importance of the need for planning, since it is based on neoliberal assumptions, excluding any form of planning for the use of natural resources that does not take into account the needs of the market.

The other concept raised by Almeida (1993) refers to eco-development. This term was introduced, according to Layargues (1997, s/p), in 1973 by Maurice Strong and "consisted of the definition of a style of development adapted to the rural areas of the Third World, based on the judicious use of local resources, without compromising the depletion of nature", since there was the possibility of these societies not being deluded by the idea of masked growth, marked mainly by social injustices. In the 1980s, Ignacy

Sachs refined this term, creating pillars of development based on economic efficiency, social justice and ecological prudence. This term is based on the premise of planning for the "ecological health of human civilization" (SACHS, 1986, p. 48 cited by FREY, 2001), with a view to conserving natural resources and the cultural characteristics of each society. In order for these objectives to be achieved, Layargues (1997) points to the need for planning aimed at new approaches, in order to benefit and conserve natural resources and the culture of peoples. In this way, Frey (2001, s/p) states that the starting point for eco-development is "a new concept of development that is compatible with the demands of the carrying capacity of the natural environment and that understands popular participation as a planning tool for achieving the proposed mode of development".

Of the two currents of thought presented, both preach development in such a way as to guarantee the preservation of nature and natural resources, i.e. ensuring the preservation of a common good for this and future generations, taking into account that all people are responsible for this common good and that they are all equal before the law. However, when analyzing the term "sustainable development", and finding that it is compatible with the assumptions of an ideology based on self-regulation of the market, Rodrigues (2006) points to a neoliberal hegemonic domination in the construction of the idea of sustainable development.

The inevitable destiny of sustainable development would be promoted by the rich countries - which have the techniques and the capital - and which would 'help' the poor countries. In other words, the continuity of development is proposed, but a word is added that doesn't change the exploitation and plundering of the workforce and natural wealth. The interest in the expanded reproduction of capital is a reaffirmed right that will provide living conditions for future generations (RODRIGUES, 2006, s/p).

In relation to this, Layargues (1997) also mentions that the idea of sustainable development foresees the installation of the total market in the economy as the solution to the environmental crisis. According to Layargues, what he hoped would be a new ecological reality is once again operating according to the logic of the market and capital. Masked under the symbol of reform, sustainable development takes on a neoliberal stance, reproducing the hegemonic ideology, as it leaves aside reflections on individual rights, class conflicts, territory, among others. This questioning is evident in the study area of this paper, since the discourse of current legislation is incompatible with the picture of degradation that is emerging.

5 Cartographic representation: application and new technologies

Based on the issues discussed throughout the paper, cartographic representation stands out as an important ally in environmental planning. Combined with information that represents the spatial situation, it is capable of helping to make decisions, as well as enabling an integrated study of the landscape, presenting and relating comprehensive information on the most varied topics that can influence environmental issues.

Thus. There is a clear need to represent the Earth's surface. It is from this need that the most varied features or phenomena occurring on the Earth's surface are highlighted. Maps are the product of this representation, and cartography is the basis for producing spatialized information. Kraak and Ormelingo (1996, p. 84) cited by Zacharias (2006, s/n), state that Cartography:

[...] provides a method or process that allows the representation of a phenomenon, or a geographical space, in such a way that its spatial structure is visualized, allowing conclusions or experiments to be inferred from this representation.

Cunha and Brito (2005, p. 3) point out the fundamental role of cartography:

[...] representing reality through spatial information in an organized and standardized way, comprising precision, determination of coordinates, mathematical resources of cartographic projections and graphic resources of symbols and texts. Its applications have been extended to all activities that require knowledge of part of the Earth's surface.

Thus, if we consider maps to be cartographic products, according to Martinelli (2003), they show much more than just the position of a particular place on the Earth's surface, as they provide detailed information about that place, characterizing it. The author also states that thematic maps should encompass three spheres of action: reading, analyzing and interpreting the information contained, presenting essential elements such as legend, title, font and scale. Zacharias (2006, s/n) adds that:

A map makes it possible to observe the locations, extensions, distribution patterns and relationships between the components distributed in space, as well as representing generalizations and extrapolations. Above all, they should encourage synthesis, objectivity, clarity of information and systematization of the elements to be represented.

The map is therefore an important tool for analyzing the interactions that take place on the Earth's surface, as it provides the possibility of reading and interpreting space in such a way as to present the most varied potentialities and/or weaknesses of a given location, which can interfere in urban and environmental planning. Driven by the need to represent the environment in which we live, man has developed various technologies capable of perfecting the making and even the interpretation of maps. Today, we are talking about Digital Cartography, which, according to Cunha and Brito (2005), brings together and integrates various types of data, resulting in maps that allow for an expanded and more complete visualization of the characteristics of a given location, since the digitalization of the most varied information makes it possible to cross-reference it, expanding the categories of analysis and prioritizing a complete understanding of the reality represented. Zacharias (2006, n.d.) states that the use of geoinformation has led to a major development in environmental mapping, since the dissemination of geographic information has benefited from the popularization of map-making techniques, leading to an increase in the quantity of maps and other cartographic information of an environmental nature. With this in mind, this article provides information on geoprocessing techniques and the interpretation of aerial photographs.

5.1 Remote sensing and photointerpretation

According to Frank (2007), remote sensing is nothing more than obtaining information about an object without having direct contact with it. Sampaio (2007) admits that remote sensing is a process of acquiring information about objects, areas or phenomena, on or near the earth's surface, by means of a sensor electromagnetic radiation placed above this surface. These sensors can be digital, installed on satellites; digital or analog, installed on planes; portable, used in field work. These sensors capture information from the Earth's surface through the energy reflected or emitted by the surface. When captured by electronic sensors, this energy is called electromagnetic radiation, the source of which can be sunlight reflected by the Earth's surface or radar (LAMPARELLI, ROCHA; BORGHI, 2001; MOREIRA, 2001; FLORENZANO, 2002; CENTENO, 2003). Sampaio (2007) points out that the objectives of remote sensing are innumerable and range from collecting and recording information to classifying and evaluating it, among others.

Thus, the use of technologies such as satellites or photographic surveys is essential for acquiring information, and the main products for obtaining information are satellite images and aerophotograms. For this research, aerial photographs were used as a remote sensing product, and the discussion prioritizes the collection of information about the technique used.

Thus, according to Affonso (2002), aerophotogrammetric surveys are carried out using special photographic cameras installed in aircraft. For Affonso, the first stage to be completed is aerophotogrammetric coverage, which is based on a flight plan. This is defined mainly according to the purpose of the project and depends on climatic and atmospheric factors for its occurrence, as well as the time of day, time of year, among others. These elements will be reflected in the quality of the photos, which will be interpreted. According to Anderson (2002) and Caldas (2006), the Iotointerpretation of a satellite image or aerophotogram is based on the identification of basic elements such as tone, texture, size, shape, shadow, standard height, location, drainage pattern, density, slope and adjacencies, and this interpretation must take place in such a way as to detect and recognize the object in the photogram so that it is possible to separate it from the rest of the set presented; also points to an analysis that goes beyond the obvious and makes it possible to systematize the information, taking into account the indispensability of fieldwork when it comes to the reliability of the final product. Caldas (2006) states that both images and aerial photographs can be interpreted visually or through automated procedures, with each form having advantages and disadvantages.

5.2 Geoprocessing

As can be seen, the increase in population, growing industrialization and disorderly occupation in urban centres, all the result of a lack of territorial planning, have led to environmental degradation. This is the context of the degraded area located on Avenida Vitalina Marcusso, in the municipality of Ourinhos/SP. Due to the magnitude of the erosion process and the risks it poses to the physical environment in which it is located, we sought to study its evolution over time, using the micro-basin of the Àgua da Veada stream as a focus for analysis. To this end, geoprocessing techniques and the interpretation of aerial photographs will be used. A discussion of the theoretical basis of these technologies was therefore considered pertinent.

According to Xavier-da-Silva and Zaidan (2004), Geoprocessing is an instrument that involves a set of concepts, methods and techniques, whose data collection is related to Remote Sensing, with the aim of carrying out analysis on the most varied nuances of spatial distribution. Xavier-daSilva (2001) cited by Caldas (2006) also states that Geoprocessing works through a set of computer procedures and operates on georeferenced databases originating from remote sensing, digital cartography or other sources, in order to carry out classifications and other transformations relating to the organization of geographic space. Veiga and Xavier-da-Silva (2004, p. 188) complete this statement when they mention that Geoprocessing enables "spatialization of information, greater accessibility, precision and speed in obtaining and processing the data needed for analysis". Thus, this tool optimizes phenomena that occur in space, integrating varied and pertinent information into a database that can always be updated, allowing for a systemic analysis of the situation.

Moura (2003) states that Geoprocessing is a term that arose from the need to process georeferenced data. According to this author, this tool has enabled progress to be made in the way the Earth's surface is represented, since it combines digital image processing, digital cartography and geographic information systems. According to Xavier-da-Silva (1999), the data transformations carried out using geoprocessing must always be done using a territorially referenced database, because only in this way will the application of the available resources produce the expected results.

Brites et al. (1998) cited by Frank (2007, p. 9) points to the importance of using geoprocessing:

Geoprocessing has become an important tool for carrying out environmental projects. The vast areas usually covered by these projects, as well as the large number of variables covered by them, make the use of geoprocessing the main resource for handling large databases.

involved in them, whether of a spatial nature or not.

It is then necessary to georeference these occurrences, associating them with a system of geographic coordinates. This process is carried out using a Geographic Information System (GIS) or Geographic Information System (GIS), software which, according to Veiga and Xavier-da-Silva (2004, p. 190), "manipulates objects (or geographic features). 190), "manipulate objects (or geographic features) and their attributes (records that make up a database) through their spatial relationship (topology)". These authors go on to add:

The geoprocessing techniques used for analysis in an IMS make it possible, for example, to define the potential of a given area for one or more activities and to combine this potential with other characteristics of these areas in order to further refine the study. The ability of an IMS to allow rapid modification, with the addition or removal of barriers, and to investigate the complex interrelationships between various thematic information plans is undoubtedly attractive for geo-planning and land management. This dynamic and interactive tool can always be readjusted as new data becomes available and as requirements and priorities change (VEIGA; XAVIER DASILVA, 2004, p. 191).

It is therefore understood that in order for a remote sensing product to be presented in the form of a map with spatialized geographic information, it must go through the georeferencing process. In this way, the image is given cartographic precision by being associated with a coordinate system. The image then takes on the characteristics of a map, as it is given properties such as scale and projection. With georeferencing, it is possible to delimit real measurements such as area, distances, among others; geographically locate any point on the Earth's surface and store relevant information for a wide variety of studies.

For the term topology, Marini (1998) quoted by Moura (2003, p. 13) explains that "the topological relationships between elements are neighborhood, pertinence, connection, inclusion, intersection; as if the computer were working with sets and identifying elements that belong to the same set according to certain characteristics". The author goes on to say that when working with data in raster format, it is easier to identify topological relationships, as there is a point-by-point image formation structure, which makes the adjacency of two pixels more easily verifiable.

Medeiros and Câmara (2001, p. 38) state that GISs are systems that process geographic data computationally. This software "stores the geometry and attributes of the data that is georeferenced (...) and its main characteristic is the diversity of generating sources and formats presented", being used in the production of maps, in the spatial analysis of phenomena and as a database. Its application is wide-ranging and can cover topics such as geophysics, urban planning and land management, among others. Medeiros and Câmara (2001, p. 39) point out that the main characteristics of GISs are:

Insert and integrate spatial information from cartographic data, census data, urban and rural cadastre, satellite images, networks and numerical terrain models into a single database; offer mechanisms for combining the various pieces of information through manipulation and analysis algorithms, as well as for consulting, retrieving, visualizing and plotting the contents of the georeferenced database.

Rosa (2003) believes that combining remote sensing techniques with GIS can provide important information on the organization of space, since data is acquired in a secure, global and periodic manner, as well as making it possible to map and gather information on the use and occupation of land in a region. In this way, geographic data would be integrated which, through the techniques presented above, would result in a cartographic product capable of enabling the integrated analysis of a series of factors that represent the real situation of a given location.

It is from the discussion presented that it becomes clear how important it is to apply new technologies that make it possible to produce maps and cartographic documents, which are extremely important when representing and interpreting environmental phenomena, These are extremely important when representing and interpreting environmental phenomena, in a way that favors the visualization of these phenomena and integrates into its discussions elements that express the relevance of this type of production when making decisions, that is, directing and carrying out actions to monitor and control various situations.

When it comes to the use and application of geoprocessing, we can take the work of Campos et al. (2004) as an example. The authors aimed to identify and quantify land use in ten micro-basins of the Capivara River basin, in the municipality of Botucatu/SP, using a database structured in the IDRISI Geographic Information System (GIS). The images used were collected by the Landsat-5 sensor in bands 3, 4 and 5 and, using supervised classification, the land use classes identified were forest, agricultural use, reforestation, pasture, exposed soil and capoeira, according to the spectral response patterns. Thus, with the analysis of the results of the ten watersheds studied, the authors show that the largest part of the area is covered by forest, grassland and reforestation (more than 40%) and that pastures form the second largest vegetation cover in the area analyzed. The study contributes important considerations regarding the rational and appropriate use of natural resources and presents data that can be used in the planning and management of the activities carried out in the micro-basins.

Dainese (2001) also made use of geoprocessing tools to study anthropogenic aggressions in an experimental basin in the Jurumim Reservoir, covering the municipalities of Itai, Paranapanema and the district of Holambra. To this end, the author divided her research into phases: recognizing the significant temporal changes in land use between 1965 and 1999; comparing image treatment processes using visual interpretation techniques and non-supervised digital classification; delimiting the existing riparian forest strip, comparing the results with the limits expressed in federal legislation. In order to draw up maps of land use in the basin and obtain drainage data and images, orbital images from the Landsat-5 satellite, planialtimetric maps and aerial photographs were used as materials, all of which were processed in the SPRING GIS. Dainese (2001) concluded that the field and savannah categories suffered a significant reduction in the period analyzed (approximately 38%), while pasture areas increased significantly, advancing about 30% of the total area set aside for permanent preservation, which has only 29% of riparian forest, sharing space with residential allotments.

In another approach, Martins et al. (2007) worked with geoprocessing to analyze the variability of the vegetation index in the urban area of the municipality of Feira de Santana/BA, correlating this information with occurrences of the disease called Visceral Leishmaniasis. To this end, the authors used satellite images obtained from the Landsat-7/ETM+ sensor, establishing a time frame from 2000 to 2003. The data was processed using Arcgis 9.1, Erdas Imagine 8.0, Surfer and Idrisi software, generating products such as vegetation maps of the urban area, climate variation over the period analyzed and the spatialization of patients infected with visceral leishmaniasis. By cross-referencing all the information, the authors obtained a map classifying the urban areas at risk in the municipality of Feira de Santana/BA. As a result, Martins et al. (2007) showed that, due to the heterogeneity of the vegetation cover in the area analyzed, climatic variables such as rainfall and temperature became of significant importance when defining risk areas and helped to delimit areas of high, medium and low risk of disease infestation. They point out that high-risk areas are five times more likely to be infected than other areas. The study is clearly an example of how geoprocessing can be used, and its association with epidemiological variables allows us to focus on disease control, since it identifies and delimits areas and risk factors, which contributes to the construction and definition of a better action plan aimed at eradicating the disease and its vectors of contagion.

Francisco Filho (2004) used geoprocessing tools to define the relationship between urban violence and socioeconomic variables and the spatial distribution of this violence in the municipality of Campinas/SP. To carry out the study, the author relied on databases such as the IBGE demographic census for the year 2000 and police occurrences for the year 2001 provided by the Campinas Civil Police Station. The data was processed using the *software* SAGA (Sistema de Anàlise Geoambiental), developed by the Federal University of Rio de Janeiro. The author analyzed the categories of crimes against the person and crimes against property and infrastructure, relating each one to the population's education levels and income conditions, generating thematic maps for each applied variable, such as the Extended Crime Potential Map and the Crime Against the Person Potential Map. In this way, it was possible to verify urban violence from a psychological, anthropological, social and economic point of view, an analysis that allowed the author to conclude that crimes against people occur mainly in areas where people with low levels of education live, while crimes against property occur preferably in regions where people with access to income and high levels of education live. The importance of geoprocessing lies in the possibility of organizing disconnected data into a structured database, generating cartographic products and providing an analysis of the behavior of violence in Campinas.

6 Case study: the area degraded by urban erosion in the Agua da Veada stream watershed

We worked in an area degraded by urban erosion located on Avenida Vitalina Marcusso, in the municipality of Ourinhos/SP (Figure 4). Four soil profiles were used as materials: a profile with technogenic characteristics, located at the head of the erosion, another collected along the erosion course and a third taken from the ravine of the Àgua da Veada stream, located downstream of the degraded area. A profile was also collected that served as a comparison standard because it retained the original physical characteristics of the Latossolos, predominant in the region, located on Rua Pascoal Henrique, outside the area of influence of the erosion. A temporal analysis of changes in the landscape of the Agua da Veada watershed was also carried out, using a map of land use and occupation, as well as a survey of legislation on environmental protection areas.

Figure 4: Area degraded by urban erosion, municipality of Ourinhos/SP.

6.1 Characterization of the areas where the soil profile samples were taken

Profile 1 is a technogenic deposit, with materials that reflect the urban expansion of the municipality of Ourinhos-SP. The samples were taken from the erosion cut located on Avenida Vitalina Marcusso, Jardim das Paineiras (Figure 5). The point sampled has the following geographical coordinates: S 22° 57' 62" and W 49° 53' 44".

Figure 5: Technogenic deposit at the head of the erosion table upstream of the Àgua da Veada stream. Profile 1 sampling point

Photo: Damasceno (2008)

The Latosol sample (profile 2) was taken from Rua Pascoal Henrique, approximately 200 m from the collection point of profile 1. The sample was taken on public land, outside the area of influence of erosion (Figure 6). This profile was sampled and used as a standard for comparison because it has edaphic characteristics close to those of the original Red Latosols, which predominate in the municipality (EMBRAPA, 1999). Its geographical coordinates are S 22° 57' 48" and W 49° 53' 22". Dense vegetation (campo sujo) prevails on the site. It is possible to identify the removal of soil, most probably for civil works. Under these conditions, the profile was exposed, which made sampling easier.

Figure 6: Latosol sample Photo: Sobral (2008)

The samples from profile 3, which characterize the changes suffered by the soil when it is exposed in the form of an erosion cut, were also taken from the urban erosion located along Avenida Vitalina Marcusso, Jardim das Paineiras (Figure 7). The point sampled has the following geographical coordinates: S 22° 57'58" and W 49° 53'40".

Figure 7. Profile sampled in the middle section of urban erosion Photo: Sobral (2008)

Soil samples were taken from profile 4, exposed in the form of a ravine at the beginning of the Agua da Veada stream, a tributary of the Pardo River, which in turn flows into the Paranapanema River. The geographic coordinates of the point sampled are: S 22°57'26"; W 49°53'18" (Figure 8).

Figure 8: Obtaining area of perii! 4

Photo: Sobral (2009)

The erosion in question is part of a private property, owned by the Marcusso family since 1935. The land was allotted by the Santa Paula company, resulting in the construction of the Jardim das Paineiras and Jardim Brilhante neighborhoods. The area was initially occupied by coffee plantations, eucalyptus trees and pastures, and nowadays it is used for allotments and pastures. In an interview with Peretti (2008), a resident of the Marcusso family who lives near the erosion site, he said that the avenue that borders the degraded area, Avenida Vitalina Marcusso, was built because of the need to access the Faculty of Technology (FATEC), which was built on land donated to the city by Jacinto Ferreira de Sà.

In the initial stretch of the avenue, in a central-to-urban direction, the launch point for the rainwater and wastewater drainage network of the adjacent neighborhoods was built: Jardim das Paineiras, Jardim Brilhante, Jardim Esplendor, Jardim Esmeralda, Diamante, Ideal, as well as the water collected in the courtyard of SESI and the Mental Health Hospital. This work was responsible for the degradation in the form of erosion in this area. The development of erosion at the site was aggravated by the inadequate sizing of the work. The erosion hotspots were approximately 600 m long, up to the source of the Agua da Veada stream. At the most critical point, it was up to 15 m wide and 9 m deep.

In a misguided attempt to contain the advanced erosion process, rubble and the remains of building materials were dumped and later buried on the site, resulting in the formation of technogenic deposits, predominantly orbic. In an interview with the then Urban Planning Coordinator of the Ourinhos Municipal Urban Development Department (RAMOS, 2007), these deposits are not known to the city council and, according to him, "constitute acts of disregard for the environment". As a result, all the deposition of the organic material has led to an alteration of the original characteristics of the red latosols that predominate in the municipality (EMBRAPA, 1999).

Under these conditions, all the material disaggregated and transported by the water, be it soil particles, roof tiles, bricks, plastic bags, household utensils, gravel, solid urban waste in general, is deposited in the Agua da Veada stream, a tributary of the Pardo river, the city's supply source and a tributary of the Paranapanema river. This process tends to further aggravate the degradation of the natural resource of water. In addition to the degradation of the surroundings, there is not even a trace of riparian forest along the stream, so there is no physical impediment to the deposition of materials and the undermining of the banks (Figure 9).

Figure 9: Siltation in the Àgua da Veada stream

Photo: Domingos (2008)

The land that backs onto the degraded area is still used as pasture covered in *Brachiària decumbens*, which shows no signs of proper management. It has around 40 head of beef cattle sold to the region's meatpackers. The zoogenic erosion caused by cattle trampling is notorious. Thus, not only at the point where the profile samples were taken, but throughout the area, there were signs of degradation of the natural resources of water, soil and vegetation.

The site underwent a "recovery" process funded by the Growth Acceleration Plan (PAC) (Figure 10), but the work was largely damaged by the rains that occurred during the period. Ourinhos City Hall has installed contour lines on the site, with the aim of slowing down the speed of the water flowing over the land, reducing the erosive capacity. The erosion is being filled in with leftover building materials.

Figure 10: PAC works in the research area Photo: Sobral (2009)

6.2 Procedures

The methodological procedures adopted for collecting soil samples were carried out in accordance with Santos et al. (2005). With regard to staking the erosion, the method used was recommended by Guerra (2005) and consists of placing stakes around the erosion area to enable monitoring.

In order to draw up the Use and Occupation Maps of the Àgua da Veada stream watershed, in the municipality of Ourinhos/SP, for the years 1972, 1984 and 2004, the methodology used was based on the principles of Photointerpretation, starting with the identification of land use and the signs of the start of the erosion process. Digital image processing software and a geographic information system (GIS) were also used to support studies on the identification of the erosion process, as well as to set up a digital database that will make it possible to handle information from current and future monitoring. For the georeferencing of the aerial photographs used, the method of collecting control points represented by UTM coordinates was used, and 8 existing control points on the ground and on the aerial photograph were identified. The data obtained in the field was then entered into the GIS Idrisi for Windows, Taiga version. With the georeferencing of aerial photographs No. 6203 from 1972, No. 33803 from 1984 and No. 019 from 2004, new images were obtained, from which it was possible to draw up the Use and Occupation Map of the Agua da Veada stream watershed, in the municipality of Ourinhos/SP. This watershed was identified using the highest elevations around the drainage area considered in the topographic map of the municipality of Ourinhos, scale 1:50,000, drawn up by the IBGE in 1973. This material was used to analyze land use and occupation in the region.

Bibliographical research was carried out on federal legal texts such as the 1988 Federal Constitution, the Cities Statute, the Forest Code, CONAMA Resolutions, as well as an analysis of the Ourinhos/SP Municipal Master Plan.

6.3 Field methods

According to Kiehl (1979), the aim of collecting and preparing soil samples is to obtain suitable material to send to analysis laboratories in order to carry out the relevant laboratory methodological procedures, ensuring the credibility of the results obtained through physical and/or chemical analysis. The aforementioned author also states the importance of rigor when collecting samples, so that they can effectively represent the study area.

Four profiles were cleaned (approximately 40 cm) at locations deemed convenient and representative for the objectives of the work: a technogenic deposit and a Latosol, measuring 100 and 150 cm with three layers and two horizons, respectively. Profile 3 was sampled along the erosion slope and profile 4 in the ravine of the Agua da Veada stream, measuring 120 and 150 cm with three horizons each, respectively. The necessary delimitations and measurements of the horizons and layers were made and approximately 1000 g of material was collected from each section for the purposes of physical analysis of each profile sampled. The procedures were carried out in accordance with Santos et al. (2005) and were carried out by the scholarship holder in the company of a support team. The physical analyses were carried out at the Geology, Geomorphology and Pedology Laboratory of UNESP/Ourinhos Experimental Campus, in accordance with the EMBRAPA soil analysis manual (1997).

With the profiles exposed and properly prepared, two *Koppec* rings were collected from each layer/horizon to determine soil density. The field activities were made possible by the use of the following materials: hoe, cutting shovel, pick, GPS, plastic bags, tape measure, pedologist's hammer, Koppec ring, castle, petri dishes, Geomorphological Map (ROSS; MOROZ, 1997) and Pedological Map (OLIVEIRA, 1999), clipboard, note-taking material, camera, etc.

6.4 Staking and erosion monitoring

Guerra (2005) states that the monitoring of an erosion process should be based mainly on systematic measurements of the study area, and that data collection can take place over a fixed period of time or not. This research is based on the monitoring of an erosion site located on Avenida Vitalina Marcusso, in the municipality of Ourinhos/SP. The aim is to follow the spatio-temporal evolution of the erosion after each episode of rain. The methodology used is recommended by Guerra (2005).

[...] stakes are placed in the ground around the gullies, about 20 meters apart (this distance can be greater if the gullies are kilometers wide), at least 10 meters from the edges of the gullies. Once these stakes have been placed, a diagram of the spatial distribution of the stakes must be drawn up and the distance from each stake to the edge of the gully measured. Every two or three months (or at a frequency related to the temporal distribution of rainfall), the site is returned to and a new measurement is taken. (GUERRA, 2005, p. 34)

It should be noted that the methodology was adapted. The erosion studied is about 700 meters long and, due to its magnitude, it was decided to keep the piles 50 meters apart.

Each stake is 10 meters from the edge of the gullies and gullies. Measurements are taken at each rainfall event or period, and the depth of the erosion is also measured. Using this technique, it is possible to trace the degree of evolution of the erosion process both spatially and temporally. In this way, 23 stakes were monitored on both sides of the erosion: stakes 1 to 14 on the left and 15 to 23 on the right, during the months of April to July 2009. Figure 11 shows the layout of the stakes along the erosion.

Figure 11: Layout of the piles along the erosion line Prepared by: Sobral (2009)

6.5 Physical analysis: laboratory procedures

The physical analysis of the four profiles sampled was carried out at the Geology,

Geomorphology and Pedology Laboratory of the Ourinhos Experimental Campus, in accordance with EMBRAPA (1997).

The following determinations were made:

- Textural analysis of TFSA: pipette method, expressed in g.kg-1. Textural class determined by the textural class distribution diagram for the granulometric classification of the material constituting the profile layers (LEMOS; SANTOS, 1984 cited by OLIVEIRA et al., 1992);
- Particle density: volumetric flask method, expressed in $kg.dm^{-3}$;
- Soil density: volumetric ring method, expressed in $kg.dm^{-3}$;
- Total soil porosity (VTP): indirect method = 100 X (DP - DS) / DP. Where: DP=
- Particle density; DS = Soil density. Result given in percentage;

7 Results and Discussions

The general description of the sampling points was made according to Santos et al. (2005), covering physical aspects of the municipality of Ourinhos, such as soil type, climate, relief, among others. Each profile sampled was also highlighted, providing information such as geographical location, local relief, current land use, among others (Table 1).

As can be seen in Table 1, the soils in the municipality of Ourinhos come from the weathering of basalt, an extrusive magmatic rock of the Sao Bento Group, Serra Geral Formation. This intense magmatic activity took place notably in the Parana Sedimentary Basin, between 147 and 119 million years ago, more specifically between the Upper Jurassic and the Lower Cretaceous of the Mesozoic Era (IPT, 1981). The original vegetation of the region was marked by the presence of the Semideciduous Seasonal Forest or Inland Atlantic Forest, which covered a large part of the state of São Paulo when it was first colonized. The initial vegetation cover was severely devastated during the process of occupation of São Paulo territory, and remnants of the forest are sheltered in state parks and isolated points in the state. The collection sites, which previously had such a plant characteristic, are now dedicated to pastoral activities and the removal of earth for construction, demonstrating the intense modification that the region has undergone over the years. This new vegetation configuration also poses a risk of particle detachment, since there is no physical impediment to the detachment, transportation and deposition of materials along the watercourse, which contributes to the silting up of the Agua da Veada stream. This gives a general description of the four points sampled during the survey (Table 1).

Table 1. General description of the soil profiles collected

Note	Technogenic deposit Profile 1	Latosol Profile 2	Middle section Profile 3	Stream gully Profile 4
Latitude	S 22° 57' 62"	S 22° 57' 48"	22° 57'58"	S 22°57'26"
Longitude	W 49° 53' 44"	W 49° 53' 22".	49° 53'40"	W 49°53'18"
Situation	Cutting a gully	Exposed profile	Exposed profile	Stream ravine
Slope	3 %	3 %	3 %	3 %
Primary vegetation	Semideciduous Seasonal Forest	Semideciduous Seasonal Forest	Semideciduous Seasonal Forest	Semideciduous Seasonal Forest
Vegetation cover on profile	Graminea	Graminea	Graminea	Graminea
Current use	Pasture	Lending area	Pasture	Pasture
Geological formation	Serra Geral	Serra Geral	Serra Geral	Serra Geral
Chronology	130 million years - Upper Tertiary	130 million years - Upper Tertiary	130 million years - Upper Tertiary	130 million years - Upper Tertiary
Source material	Basalt	Basalt	Basalt	Basalt
Local relief	Gently wavy	Gently wavy	Gently wavy	Gently wavy
Regional relief	Gently wavy	Gently wavy	Gently wavy	Gently wavy
Erosion	Ravines/voçorocas	Not apparent	Ravines/voçorocas	Laminate/ravine
Drainage	Well drained	Well drained	Well drained	Well drained
Weather	Transition from Tropical to Subtropical Climate	Transition from Tropical to Subtropical Climate	Transition from Tropical to Subtropical Climate	Transition from Tropical to Subtropical Climate
Other observations	Presence of leftover construction materials, solid urban waste, storm water runoff.	Lending area for construction materials, possibly	Presence of leftover construction materials, solid urban waste, storm water runoff.	Stream bank showing signs of degradation, as well as the presence of a layer of sediment in the profile.

The water erosion present at the site where profiles 1 and 3 were sampled resulted from the concentration of rainwater coming from the drainage network located on the site, a public and poorly dimensioned work. Profile 4 was also directly influenced by this work, which led to the existence of a surface layer of sediment and particles carried to the lowest point of the terrain.

7.1 Physical analysis of the soil

Tables 2, 3 and 4 show the data from the routine physical analyses carried out at the Geology, Geomorphology and Pedology Laboratory at UNESP/Ourinhos Experimental Campus. The following properties were analyzed: texture, soil density, particle density and porosity. We also worked on scraping the profile and obtaining a homogeneous sample for each profile sampled in order to quantify and qualify the technogenic materials.

Table 2. Results of the determination of soil density, particle density and porosity.

Samples	Thickness (cm)	Soil density ($kg.dm^{-3}$)	Particle density($kg.dm^{-3}$)	VTP (%)
Profile 1				
Layer 1	0 - 35	1,60	2,70	41
Layer 2	35- 65	1,66	2,60	36
Layer 3	65 - 100+	1,58	2,60	39
Profile 2				
Horizon A	0 - 70	1,30	2,32	43
Horizon B	70 - 150+	1,40	2,32	39
Profile 3				
Layer 1	0 - 40	1,37	2,61	47
Layer 2	40 - 70	1,32	2,92	55
Layer 3	70 - 120+	1,29	2,85	55
Profile 4				
Layer 1	0 - 20	1,56	2,47	37
Layer 2	20- 50	1,63	2,65	38
Layer 3	50 - 150+	1,77	3,18	44

* Total Pore Volume

With reference to the data presented in Table 2, it can be seen that: **Profile 1**: shows a process of compaction manifested in the soil density values, possibly associated with cattle trampling and the great diversity of technogenic materials. The values found are in line with those presented in the literature on urban soils. Under these conditions, the higher the soil density, the greater the compaction, the greater the impairment of air circulation and water percolation, and consequently the greater the vulnerability to water erosion.

Regarding the determination of particle density, through the results of the samples from profile 1 (Table 2), given the fact that this is a deposit with extremely heterogeneous materials, it is only possible to state that the material analyzed is eminently mineral.

In layer 1 (profile 1), although configured as a technogenic deposit, the porosity of 41 % is in line with the limits expected for clayey soils such as the Latossolos in the region. The variation in porosity values at depth in profile 1 can also be attributed to the diversity of materials. Under these conditions, theoretically there would be no major compromises in terms of water percolation and air circulation.

Profile 2: shows a lower degree of compaction compared to profile 1. This behavior was expected since this profile is in conditions of less anthropogenic intervention. The A horizon has a lower density than the B horizon, 1.3 and 1.4 $kg.dm^{-3}$, respectively. This may be associated with the greater amount of organic matter on the surface, which favors a reduction in soil density. The increase in soil density at depth, in the B horizon, can be attributed to the weight of the overlying layers.

According to the determination of particle density, profile 2 is made up of predominantly mineral material. Even the A horizon shows this behavior, since the vegetation cover is not sufficient to form an organic horizon, which is the necessary condition for low particle density indices.

With regard to VTP, there is a positive correlation between an increase in soil density (B horizon; profile 2) and a decrease in porosity.

It should be noted that, although the site had suitable conditions to be used as a benchmark, it is located in an urban context and is therefore subject to the processes of soil formation: addition, transportation, transformation and loss.

Profile 3: the average soil density result is 1.32 kg. dm^{-} . This density value is considered low and is representative of sandy mineral soils. This condition does not pose any difficulties for plant growth as root penetration is not hindered. In other words, although the soil profile in question is located in an urban environment, its apparent density is considered good, as it does not exceed 1.4 kg. dm^{-} . With regard to determining particle density, the results shown in Table 2 make it possible to characterize the soil as mineral, with the mineral components being considered light, since the average particle density is 2.79 kg. dm (KIEHL, 1979). The porosity factor is considered acceptable according to the standards set for clay soils, which according to Kiehl (1979) range from 40% to 60% pore size. Based on these results, profile 3 was theoretically not compromised in terms of the amount of existing pores, which did not hinder the percolation of water in the soil. The variation in porosity in the horizons can be attributed to the de-characterization suffered by the soil due to its exposure in erosion cuts.

Profile 4: this profile is characteristic of a stream bank, suffering iron reduction processes due to the presence of water. As a result, the soil sampled has an average density of 1.65 kg.dm^{-} , indicating severe compaction, which may be due to cattle trampling and the various interventions suffered by the soil at the site. The density values increase as the profile deepens, which is acceptable as this is due to the weight of the overlying layers. However, the high density of the soil is an impediment to the growth of plant roots. In addition, it is possible to observe the accumulation of sediment carried by rainwater in the surface horizon of the profile. As for particle density, in general this profile can be characterized as mineral, with an average value of 2.76 kg.dm^{-}. With regard to VTP, there is a positive correlation between an increase in soil density and a decrease in porosity. However, this porosity is representative of sandy soils, where the percentage of pores varies between 35% and 50% (KIEHL, 1979).

To the results obtained for the four profiles analyzed, we must add the information that they are located in an urban context, subject to modification and alteration due to the advance of urbanization, dumping of domestic waste, removal of forest, among other factors that determine the de-characterization of soil classes in general, resulting in results that are incompatible with the predominant soil class, Latossolos.

From a textural point of view (Table 3), profile 1 shows an increase in what would be the sand fraction in layer 2 (733 g.kg^{-}), as can be seen in Table 3. If this were a Latosol from basalt, this behavior would be atypical, since this class of soil shows gradual or diffuse textural change. Under these conditions, the addition of materials other than the source rock is once again evident. This statement is also based on the fractionation of the supposed sand. The presence of very coarse fractions in the samples from profile 1 does not correspond to the characteristics of the source material. Therefore, the formation of this profile is related to the dynamics of society, not nature.

Homogenizing the samples from the technogenic deposit (profile 1), it can be seen that the results for sand, clay and site are close to those expected for a Latosol, despite showing traces of the very coarse fraction.

Profile 2, considered a standard because it apparently has characteristics close to Latosols, does not show any textural variation along the profile that suggests an abrupt textural change (Table 3). This confirms the gradual or diffuse change in horizons typical of latosols. Under these conditions, water percolation would not be compromised, since there are no impeding horizons.

From a textural point of view, profile 3 shows a considerable amount of sand in all the horizons observed, which is incompatible with the predominant soil class in the municipality, Latosol. This result shows that the soil has been altered and is incompatible with its original material. However, a greater amount of sand was observed, represented mainly by fine to very fine sands, a fact that refers to the current situation of the research area: highly degraded by water erosion and intensely mobilized due to soil loss by erosion. This result is reflected in the infiltration capacity of the profile, which suffers a process of differential erosion due to the size of the particles.

Table 3. Textural analysis results

				Sand fractionation*				
Samples	Sand	Silt	Clay	MG	G	M	F	MF
Profile 1								
Layer 1	573	123	304	1	5	34	40	20
Layer 2	733	123	144	1	5	36	40	18
Layer 3	544	138	308	7	15	34	31	13
Homogeneous	546	144	300	3	9	34	37	17

Profile 2								
Horizon A	587	99	314	0	4	34	39	23
Horizon B	592	102	306	0	5	33	40	22
Homogeneous	591	119	290	0	6	34	38	22
Profile 3								
Layer 1	634	206	160	0	0,6	7,6	47,5	44,3
Layer 2	646	101	253	0	0,8	8,4	67,4	23,4
Layer 3	648	192	160	0	0,6	7,8	47	44,6
Homogeneous	642	132	223	0	0,3	6,3	44,4	49
Profile 4								
Layer 1	655	185	160	0	0,3	3,3	20	76,4
Layer 2	327	247	426	0	0,6	7,6	47,5	44,3
Layer 3	721	119	160	0	0,6	5,2	38,8	55,4
Homogeneous	641	199	160	0	0,3	5	35,6	59,1

MG: very coarse sand; G: coarse sand; M: medium sand; F: fine sand; MF: very fine sand

It is important to note that profile 4 is located in a valley bottom area, which suffers from the action of water and the consequent reduction of iron, resulting in a soil with hydromorphic characteristics, as well as the B horizon being an impediment due to the large amount of clay it contains (Table 3). When the sand was fractionated, it showed a large amount of fine to very fine sand, a characteristic that dates back to the source material. The absence or insignificance of coarse elements is consistent with the characteristics of the soil found at the site.

Comparing the last two profiles sampled with the Latosol profile, taken as a standard and adopted as a comparison parameter, we find that:

Profile 3: when compared to profile 2, Latosol, profile 3 has a slightly lower degree of compaction than the former, with the values found being 1.32 kg.dm^{-} and 1.35 kg.dm^{-}, respectively. This means that both profiles are not difficult for roots to penetrate or for air and water to circulate. Another factor that should be compared is particle density, whose values established for the two profiles represent mineral characteristics. The porosity of profile 3 showed a positive correlation with soil density: lower density, higher porosity. This may be related to soil management in both profiles, with the former showing great instability; however, the values found for profile 3 are acceptable when dealing with clay soils. As far as textural analysis is concerned, profile 3 shows a significant amount of sand when compared to profile 2, which is evidence of the strong alteration of the physical characteristics of these typically clayey latosols.

Profile 4: this profile is in the context of a stream and has suffered iron reduction due to the presence of water. It is therefore a hydromorphic soil. When compared to profile 2, it does not have similar soil density characteristics, since profile 4 is more compacted than profile 2, reaching an average of 1.65 kg.dm , making it difficult for roots to penetrate, air to circulate and water to percolate, which favors water erosion. With regard to particle density, both profiles show mineral characteristics. The porosity factor behaves differently when the two profiles are compared. Profile 3's porosity factor increases as soil density increases; profile 2's porosity factor decreases as soil density increases. This factor is therefore directly and inversely proportional to soil density in profiles 3 and 2, respectively. This finding is related to the type of use applied to each profile. When compared in relation to the textural analysis, profile 4 was found to have a impediment horizon, which would make it difficult for water to percolate throughout the profile; profile 2*, on the other hand,* does not have this type of formation, so there is no impediment to water percolation in the profile, which is typical of latosols.

Considering the fact that the profiles analyzed were inserted in a context of urban degradation, evidenced by the presence of gullies and gullies, it is understood that these values are altered since their area of occurrence is marked by anthropic intervention. Profile 4 is typically characterized as hydromorphic and also receives direct influence from technogenic deposits, as it is located downstream from the erosion, receiving sediments and materials that form an atypical surface horizon in pedological formations of this type.

The materials found in the technogenic deposit (profile1) shown in Table 4 were also quantified and qualified.

Table 4. Quantitative and qualitative analysis of the materials identified in the technogenic deposit profile

		Quantification of materials (%)			
Profile 1	**Thickness (cm)**	**Sieve opening (mm)**			**Urban materials (bricks, tiles, gravel, PVC pipes, concrete, tiles, etc.)**
		>10	>10 - 4	4 -2	**(%)**
Layer 1	0-35	49	12	11	61
Layer 2	35-65	55	17	13	67
Layer 3	65-100+	44	24	12	15

The artifacts found in the technogenic deposit (profile 1) were selected by sieving from a 1000 g sample obtained from each layer (Figure 12).

Figure 12. Selection by sieving of samples from the technogenic deposit. Photo: Sobral (2009)

As can be seen in Table 4, there is a predominance of uric materials in the first two layers, which confirms and reaffirms the existence of the technogenic deposit at the site. The third layer shows less incorporation of materials, which can be explained by the depth at which this layer is located. However, in all the layers analyzed, it is possible to see the presence of artifacts whose dimensions exceed 10.0 mm, since there are remains of construction materials (layers 1 and 2). This information was to be expected, since the urbanization process of the surrounding neighbourhoods, which began around 20 years ago, is characterized by masonry houses, as well as the asphalting of the streets and the population that ends up depositing household and construction waste on the site. Over time, materials have been incorporated into the soil system, altering its physical properties.

7.2 Erosion monitoring

It should be borne in mind that the area analyzed is highly degraded by water erosion. Cattle trampling and poorly managed grazing, together with the channeling of rainwater from the surrounding neighborhoods, are the main triggers of erosion. Table 5 shows the soil loss per rainfall event.

Table 5: Advance of urban erosion by event or rainy season from April to July 2009 (meters)

Rainfall/ Stakes	13 e 14/04	27/04	02 e 03/05	08/05	24/06 27,06	a01/07	04/07	Total
1	0	0,06	0	3,3	0,7	0,1	0,1	4,26
2	0	0	0	0	0	0	0	0
3	0	0	0	0	0	0	0	0
4	0	0,02	0	0	0,1	0,1	0	0,4
5	0	0	0	0	0	0	0	0
6	0	0	0	0	0	0,2	0	0,2
7	0	0	0	0	0	0	0	0
8	0	0	0	0	0	0	0	0
9	0	0	0	0,1	0	0	0	0,1
10	0	0	0	0	0,1	0,2	0	0,3
11	0	0	0	0	0	0	0	0
12	0	0	0	0,4	0	1,6	0	2
13	0	0	0	0	0	0	0	1,4
14	0,05	0	0	1,25	0,1	0	0	0
15					0	0	0	0

16	0	0	0	0
17	0	0	0	0
18	0	0	0	0
19	0	0	0	0
20	0	0	0	0
21	0	0	0	0
22	0	0	0	0
23	0,8	0	0	0,8

Table 6: Depth of the points monitored within the erosion at each rainfall event (meters)

Rainfall/ Stakes	13 e 14/04	27/04	02 e 03/05	08/05	24/06 a 27,06	01/07	04/07
1	2,0	2,0	2,0	2,0	2,30	3,20	3,20
2	2,45	3,30	3,0	2,45	3,40	3,33	3,33
3	3,27	3,20	3,20	3,20	3,30	3,30	3,30
4	2,0	2,10	1,96	1,96	2,10	2,10	2,10
5	2,27	2,20	2,30	2,15	2,20	2,70	2,70
6	4,60	4,50	5,0	5,0	5,0	5,50	5,50
7	8,30	8,10	8,60	8,60	8,10	8,10	8,10
8	7,40	7,0	7,0	6,50	7,30	7,30	7,30
9	3,70	3,70	3,70	3,70	6,50	6,50	6,50
10	1,10	1,20	3,80	3,80	3,0	3,10	3,10
11	1,20	1,0	0,70	0,70	1,30	1,30	1,30
12	6,30	6,30	8,20	8,20	6,70	3,70	3,70
13	3,40	4,30	6,0	6,0	6,30	6,30	6,30
14	6,45	5,50	3,10	3,10	3,20	3,20	3,20
15					3,0	3,0	3,0
16					3,70	3,70	3,70
17					0,30	0,30	0,30
18					2,0	2,0	2,0
19					1,10	1,10	1,10
20					4,70	4,70	4,70
21					3,60	3,60	3,60
22					2,70	2,70	2,70
23					3,10	3,10	3,20

From the analysis of Tables 5 and 6, it is possible to see the loss of soil due to water erosion at some of the monitored points, which is increasing and expressive. When it comes to depth, this information is somewhat irregular, since the materials carried by the rains can be removed from the system and are often deposited in the watercourses, resulting in a reduction in depth. On the other hand, the process of undermining the eroded banks can cause the creek to be buried, reducing its depth.

Table 5 shows that there are points that suffered significant soil losses (stakes 1, 12 and 14). Stake 1 shows this behavior because it is the head of a ravine, where the slope is undermined due to its proximity to the launch point of the drainage network that existed on the site. Piles 12 and 14 were located in the region of the source of the stream, suffering direct action from both surface and subsurface water, and were characterized as areas of instability that should be permanently preserved.

In order to emphasize the relationship between the advance of gullies and rainfall, the rainfall indices for the monitoring period (April to July 2009) provided by INMET (2009) were taken into account, as shown in Figure 13.

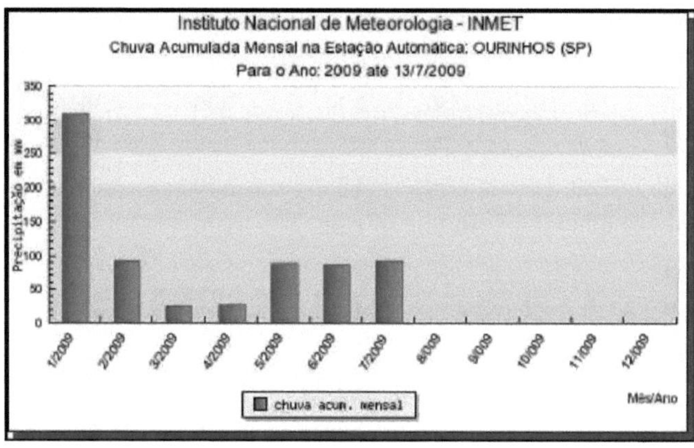

Figure 13: Monthly rainfall totals (mm) for 2009 in the municipality of Ourinhos/SP. Source: INMET (2009)

Although the data shown in Figure 9 does not represent rainfall erosivity, it is possible to draw a relationship between this variable and soil loss due to water erosion. By analyzing this figure and the data obtained through systematic field measurements, it was possible to draw a correlation between these data. Table 5 shows that the months with the highest soil losses are May and June, when the average rainfall is also high for both months. The higher the rainfall, the greater the soil loss. In addition, the texture of the soil, which has been modified due to the rains and the large amount of mass mobilization, is also responsible for the fragility and instability of the site, with the undermining of bases and the displacement of slopes being facilitated when the rains occur.

7.3 Considerations on the Master Plan of the municipality of Ourinhos/SP and the relevant legislation

Taking the discussion to the context of the municipality of Ourinhos, more specifically the degraded area of Avenida Vitalina Marcusso, the site has undergone several changes over the years, the result of a history of use and occupation without proper management techniques. It is therefore understood that this area, which is predominantly degraded pasture, is in an irregular situation with regard to the relevant environmental legislation. An example of this is the lack of riparian forest around the springs of the Agua da Veada stream, which are now completely buried by the construction of rainwater galleries funded by the PAC (Figure 14). The Forestry Code guarantees the presence of vegetation in these areas, and the Municipal Master Plan corroborates the Code's proposals. It is clear, then, that there is a complete failure to comply with the law.

Figure 14: Urban erosion recovery work on Avenida Vitalina Marcusso Photo: Sobral (2010)

Another point worth highlighting is the fact that a private area receives investment from the public authorities. Although Article 170 of the Federal Constitution guarantees the right to private property, the way in which the area was intervened is questionable. According to the Federal Constitution in its article 182, paragraph 2, it is clear that private urban property fulfills its social function by meeting the requirements of the Master Plan. For the intervention carried out in the degraded area of the Agua da Veada stream, it would be necessary to expropriate or expropriate the property, since the case would be of public utility. This determination is laid down in Article 5 of Decree-Law No. 3.365/41, which states that the Public Authority has the right to acquire the dominion and use of the property of others. Only in this way could the intervention take place, as it would not benefit just one person, but the community.

It is clear, then, that the Ourinhos City Council is violating the principle of legality, laid down in Article 37 of the Federal Constitution, which states that actions can only take place in accordance with the relevant legislation and that the Public Administration has limits on its actions and must obey the law in its entirety.

7.4 Land use maps for the years 1972, 1984 and 2004 in the micro-basin of the Agua da Veada stream

It is understood that inadequate environmental and physical-territorial planning in the research area has led to urban erosion and consequently to the silting up of the Agua da Veada stream. As an important planning tool, cartographic representation, combined with information that represents the spatial situation of a given location, is able to help in decision-making, as well as enabling an integrated study of the landscape. In order to show how the landscape has changed over time, as well as the uses attributed to it, maps of land use and occupation were drawn up in the Agua da Veada stream watershed, with the aim of understanding the various activities that have taken place there and how they could have contributed to the formation of the current erosion situation. In this way, aerial photos were used at a scale of 1:25.00, 1:30.000 and 1:30.000, respectively, taken in three different periods: 1972, 1984 and 2004, using the principles of Photointerpretation and Geoprocessing.

As can be seen in Figure 15, in 1972 the Raposo Tavares Highway was already an important access route for agricultural production to the state of Sao Paulo, and, as it cut through the city of Ourinhos, it allowed urbanization to grow and advance. In that year, the research area, despite being influenced by the highway, was not very urbanized, with agricultural and livestock activities predominating. There are also secondary access roads in the watershed, linking rural properties to the city of Ourinhos, allowing the population to move around.

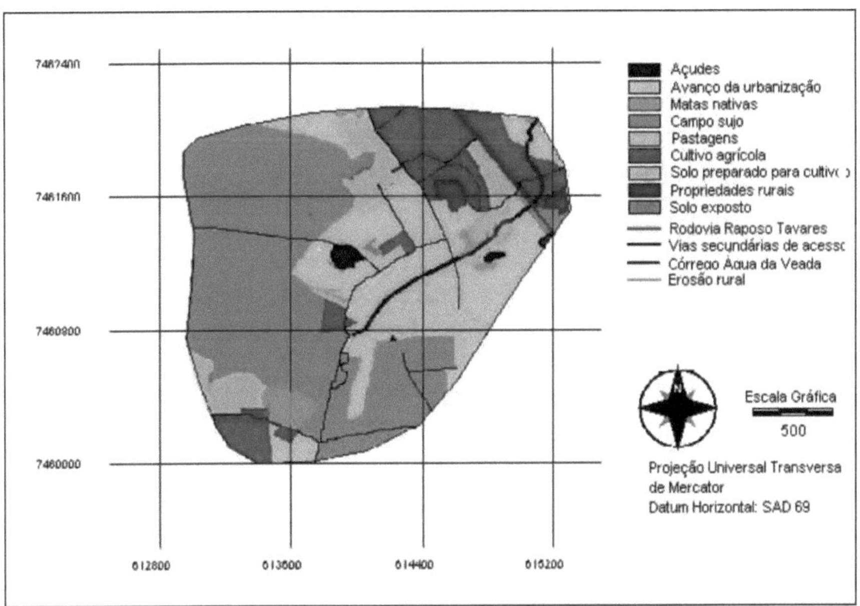

Figure 15. Map of land use and cover in the Agua da Veada stream watershed in 1972, Ourinhos/SP municipality.

Elaboration: Sobral and Piroli (2010)

Rural properties, which have been identified as having some kind of civil construction, were negligible in the watershed in 1972. As for the advance of urbanization, it is incipient, with the presence of a housing development, currently Vila Sao Luiz. The city was also growing towards the banks of the Agua da Veada stream. The watershed was then predominantly agricultural. Pastures lined the entire length of the stream, resulting in the suppression of riparian vegetation as early as the 1970s.

There are few patches of native vegetation in the area where the drainage network is now located, which is responsible for the development of erosion. There is no riparian vegetation even at the source of the stream, a fact that shows a lack of concern for the quality of this watercourse. There is an area of marshland along the stream and some dams that would have served as cattle watering holes. A vast region covered in sparse forest was recognized and characterized as a dirty field, bordered by pastures and some plantation areas, which would culminate in its later suppression. Small areas of exposed soil were also identified, sometimes marked by their insertion in pasture areas and sometimes in cultivated areas. No signs of urban erosion were identified at the time. However, rural erosion was identified, probably as a result of the lack of vegetation and animal trampling, as well as the proximity of a rural road, which may have had some influence on this manifestation.

Looking at Figure 16, it can be seen that the main activity carried out on the site is livestock farming, represented by "pasture", which makes the Àgua da Veada stream watershed still predominantly agricultural in 1984.

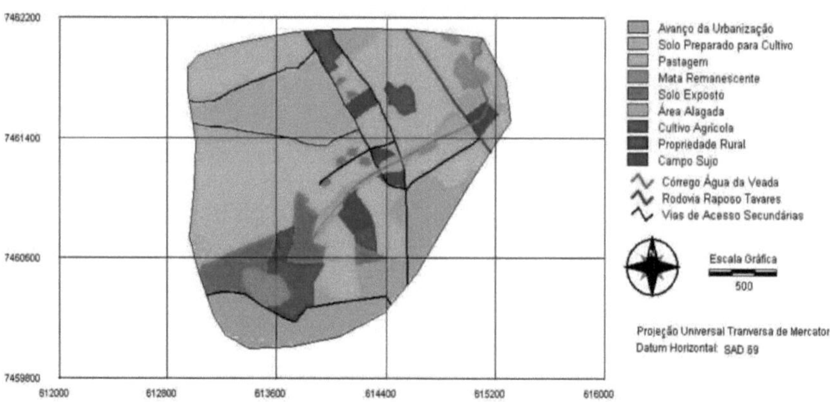

Figure 16. Map of land use and cover in the Agua da Veada stream watershed in 1984, Ourinhos/SP municipality.

Elaboration: Sobral (2010)

The figure shows that most of the riparian vegetation has been removed, making room for livestock farming. At the time the photo was taken, there were also sites for agricultural cultivation, represented by exposed soil, soil prepared for cultivation and agricultural crops. Even so, these activities are still considered secondary when compared to the areas destined for pasture. In this way, pastures still accompany the entire length of the watercourse, flanking areas of meadow and perpetuating the condition of the absence of riparian vegetation in favor of agricultural practices. There are very few patches of "remnant forest" along the watercourse, with a complete absence mainly in the region of the source and mouth of the stream. Areas of campo sujo have also been heavily suppressed, giving way to pasture. These facts confirm the lack of commitment on the part of both the municipal authorities and the landowners in this watershed, which today results in the total destruction of this resource, mainly due to silting.

With reference to the Raposo Tavares highway and its surroundings, it can be seen that in 1984 the main activity was livestock farming, which indicates the strong rural influence along the highway. As for the advance of urbanization, it is *already* more expressive, especially towards the banks of the stream, with part of the current Jardim das Paineiras and Jardim Santa Fé neighbourhoods. Secondary access roads still exist, albeit in smaller numbers, linking rural and urban areas. In relation to the rural erosion identified on the land use and occupation map in 1972, it disappeared in 1984, which suggests that the site has probably recovered. Even with the advance of urbanization, the construction of new neighbourhoods and the strong agricultural use of the micro-basin, it was not possible to identify signs of erosion.

Based on Figure 17, it can be said that urbanization is advancing continuously and expressively towards the banks of the Àgua da Veada stream, causing intense degradation, recognized by the absence of riparian forest. It can also be seen that the Raposo Tavares highway has had its banks completely taken over by urbanization, replacing pastures. There are remnants of forest near the urbanized areas, which presupposes some kind of reforestation that has taken place due to the presence of housing on the site. There has also been some increase in this forest near the mouth of the stream, which points to the weight of environmental legislation with regard to the delimitation of permanent preservation areas. However, this legislation is not applied along the stream, which is flanked by pastures and the city.

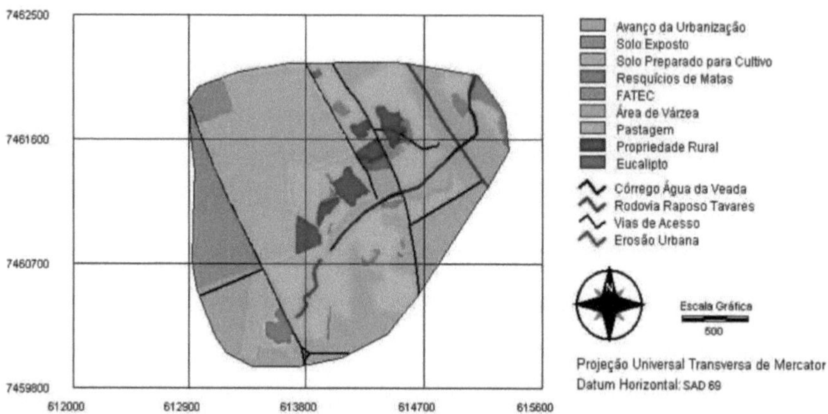

Figure 17. Map of land use and cover in the Agua da Veada stream watershed in 2004, Ourinhos/SP municipality.

Elaboration: Sobral (2010)

The figure shows human occupation in places characterized as wetlands, resulting in the vulnerability of the stream and its current configuration. Vitalina Marcusso Avenue also provides access to the Faculty of Technology (FATEC). Along this access road, we can see extensive areas destined for agricultural cultivation and pasture, as well as an increase in rural properties, identified by some type of civil construction.

On this map, it is already possible to identify urban erosion, the acceleration of which is due to the construction of a launch point for the drainage network of adjacent neighborhoods. This erosion contributes directly and constantly to the siltation of the Agua da Veada stream, as well as compromising the quality of the soil in the area, making it unsuitable for cultivation, allotments, among other things, as well as degrading the area from a public health point of view, since solid waste is dumped inside the erosion, which contributes to the formation of technogenic deposits. The agricultural character of the watershed is now permeated by the advance of the urban fabric, bringing positive and negative consequences from a socio-environmental point of view.

8 Final considerations

From the analysis of the four profiles, it is possible to see the degree of degradation that exists in the study area, as evidenced by the physical analyses carried out. What should be highlighted is the deconfiguration of the physical characteristics (technogenic deposit and middle section of erosion) due to the insertion of anthropogenic materials. This change can be seen when the profiles sampled in the erosion are compared with the Latossolo, which is considered to be closer to the original characteristics of the soils in the municipality of Ourinhos/SP. In addition, the deposition of materials and sediments along the banks of the stream stands out, forming an atypical surface horizon in hydromorphic soil formations. Profiles 1 and 3 (technogenic deposit and middle section) show characteristics such as the addition of anthropogenic materials and changes in the textural composition of the soil, which results in greater instability of the terrain, favoring the formation of subsidence steps. Technogenic deposits therefore contribute to the evolution of erosion processes.

When analyzing the data on erosion monitoring, it can be seen that it evolves due to the rainfall during the period, with the headwaters of the erosion and the source of the Agua da Veada stream standing out.

With the reality of the PAC works in the municipality of Ourinhos, more specifically in the research area, there are once again human alterations and interventions. However, these works are intended to "recover" the site. What can be seen is that part of the work has been compromised due to the interruption of work and exposure to heavy rains and, consequently, the misuse of public money.

It can therefore be concluded that the appropriation of natural resources by anthropic action results in alterations to their original characteristics. From a pedological point of view, these alterations can exceed tolerable limits, compromising the land's capacity for use. The results of the research highlight public negligence in the conservation of natural resources such as soil, vegetation and water, manifested in one of the most spectacular forms of soil degradation: accelerated erosion. This situation jeopardizes the use of the area and brings a picture of instability to the place, which generates danger for the population, since there is the deposit of solid urban waste, which can culminate in the transmission of diseases.

By analyzing the maps, it can be seen that the Agua da Veada stream watershed has suffered from the absence of riparian forest and native vegetation since the early 1970s. A large part of the watershed is used for farming, and rural erosion can be identified as a result of these activities, which are carried out without management and conservation policies. It was noticed that at this time there were no signs of urban erosion, since urbanization was incipient and the point where the drainage network was launched was a reality in the 1990s. Another fact that should be borne in mind is that, over time, the basin has become waterproofed, as several neighborhoods have been built, reducing the area that used to help infiltrate rainwater. This waterproofing of the micro-basin may also have aggravated the process of accelerated erosion development, as it contributed to larger areas that channel rainwater to the lower elevations of the micro-basin in question. The exacerbated process of modifying the landscape, such as the establishment of accelerated erosion and the accumulation of solid urban waste, especially on the outskirts of cities, tends to compromise people's quality of life.

In conclusion, it can be seen that all the intervention in the Agua da Veada stream has taken place in a way that fails to comply with current legislation, whether at federal level (Forest Code, City Statute, Federal Constitution, CONAMA Resolutions) or at municipal level (Master Plan). This non-compliance has been a reality since 1970, as can be seen in the map of the use and coverage of the watershed, highlighting the inexpressive vegetation cover, which certainly contributes to the degradation of soil/vegetation/water natural resources. Furthermore, an analysis of the maps showed that the urban area in the municipality of Ourinhos/SP is continually expanding along the banks of the Àgua da Veada stream, which calls into question the municipality's urban expansion policy, since it is taking place indiscriminately and without respecting important resolutions aimed at protecting areas of permanent preservation.

The panorama presented is incompatible with the "political" proposals embodied in the Master Plan of the municipality of Ourinhos, as well as disrespecting environmental laws, in addition to demonstrating the importance of environmental planning. This is the possible contribution of scientific research: diagnosing, sizing up and planning.

9 References

ALMEIDA, J. R. de, et al. **Planejamento Ambiental**: caminho para participaçao popular e gestão ambiental para nosso futuro comum; uma necessidade, um desafio. Rio de Janeiro: Thex, 1993.

ANDERSON, P. **Principles of topographic cartography**. Illinois State University. 2002.

ANTUNES, P. de B. **Politica nacional do meio ambiente - PNMA**: Comentários à Lei 6.938, de 31 de agosto de 1981. Rio de Janeiro: Lumen Juris, 2005. p. 16.

ARAÙJO, S. M. V. G. Permanent preservation areas and the urban question.

Legislative Consultancy. August 2002. Available at: <http://www.mp.ba.gov.br>. Accessed on: January 20, 2010.

AZEVEDO, A.C. **Soils and the environment**: An introduction. Santa Maria: Pallotti, 2006.

BENJAMIN, A. H. de V. (coord). Environmental damage: prevention, reparation and repression. Sao Paulo: **Revista dos Tribunais,** 1993.

BERTONI, J.; LOMBARDI NETO, F. **Conservação do solo.** 5 ed. Sao Paulo: icone, 2005.

BIGARELLA, J. J.; BECKER, R. D.; SANTOS, G. F. **Estrutura e origem das paisagens tropicais e subtropicais**. 2. Ed. V. 3. Florianópolis: UFSC, 2007.

BRAGA, R.; CARVALHO, P. F. C. **Recursos hidricos e planejamento urbano e regional.**

Rio Claro: Municipal Planning Laboratory - IGCE - UNESP. 2003. p. 113127

BRAZIL. Law No. 7.803, of July 18, 1989. **Amends the wording of Law No. 4.771, of September 15, 1965, and repeals Laws Nos. 6.535, of June 15, 1978, and 7.511, of July 7, 1986**. Brasilia: Federal Senate, 1989. Available at: <www.planalto.gov.br/ccivil_03/LEIS/L7803.htm#art2>. Accessed on: April 21, 2009.

BRAZIL. **Constitution of the Federative Republic of Brazil**. 05/10/1988. Available at: <http://www.planalto.gov.br/ccivil_03/constituicao/constitui%C3%A7ao.htm>. Accessed on: January 20, 2010.

BRAZIL. **Forest Code**. Law n° 4.471 of September 15, 1965. Available at: <http://www.planalto.gov.br/ccivil_03/LEIS/L4771.htm>. Accessed on: January 15, 2010.

BRAZIL. **Decree-Law** No. 6.398, of October 31, 1981. Establishes the guidelines for the National Environmental Policy. Available at: <http://www.planalto.gov.br/ccivil_03/Leis/L6938.htm>. Accessed on: January 20, 2010.

BRAZIL. **City Statute**. Law No. 10.257 of July 10, 2001. Available at: <http://www.planalto.gov.br/.../LEIS/LEIS.../L10257.htm>. Accessed on: January 15, 2010.

BRAZIL. **Ministry of the Environment**. Available at: <http://www.mma.gov.br/sitio/>. Accessed on: January 15, 2010.

CAMPOS, S., et al. Remote sensing and geoprocessing applied to land use in watersheds, Botucatu - SP. **Eng. Agric**., Botucatu, v. 24, n. 2, Aug. 2004. Available from <http://www.scielo.br/scielo.php?script=sci_arttext&pid=S0100-69162004000200023&lng=en&nrm=iso>. Accessed on: April 21, 2011.

CALDAS, A. J. F. da S. **Geoprocessing and environmental analysis to determine habitat corridors in Serra da Concórdia, Vale do Paraiba - RJ**. 2007. 110 f Dissertation (Master of Science), Institute of Forestry - Federal Rural University of Rio de Janeiro. Available at: <www.ufrrj.br/institutos/if/lmbh/pdf/mono_disset_tese27.pdf>. Accessed on: 20 de January 2010.

CAPELLI, Silvia. The environmental impact study in Brazilian reality. **Legal Studies**. Sao Leopoldo. v. 27. n. 70. p. 49-64. mai./ago. 1994. Available at: <www.mp.rs.gov.br/biblioteca/revistamp?letra=C>. Accessed on: January 25, 2010.

CARVALHO, I. C.M. **Os mitos do desenvolvimento sustentàvel**. p.17-21, nov.- dez. 1991.

CARVALHO, P. F. de; BRAGA, R. (eds.) **Perspectivas** de **Gestao Ambiental em Cidades Médias**. Rio Claro: LPM-UNESP, 2001. pp. 95 to 109.

CARRIJO, O.A. et al. Fertigation trends and challenges in Brazil. In: FOLEGATTI,

M.V. (Coord.) **Fertirrigaçao**: citrus, flowers, vegetables, Guaiba: Agropecuâria, 1999. p.155- 169.

CHAVES, J. C. D.; CALEGARI, A. Green manuring and crop rotation. **Informe Agropecuàrio**, Belo Horizonte, v. 22, n. 212, p. 53-60, Sept./Oct. 2001.

CHRISTOFOLETTI, Antonio. **Modeling environmental systems**. Sao Paulo: Edgard Blücher Ltda, 1999.

COELHO NETO, A. L. Slope hydrology at the interface with geomorphology. In: GUERRA, A. J. T., CUNHA, S. B. **Geomorfologia, uma actualizaçao de bases e conceitos.** Rio de Janeiro: Bertrand Brasil, 2003, 3 ed.

CONAMA. Resolution 001 of January 23, 1986. Available at: <http://www.mma.gov.br/port/conama/res/res86/res0186.html>. Accessed on: January 25, 2010.

CONAMA. Resolution 011 of March 18, 1986. Available at: <http://www.lei.adv.br/011-86.htm. Accessed on: January 25, 2010>. Accessed on: January 30, 2010.

CONAMA. Resolution 005 of August 6, 1987. Available at: <http://www.mma.gov.br/port/conama/res/res87/res0587.html>. Accessed on: January 30, 2010.

CONAMA. Resolution 237 of December 19, 1997. Available at:

<http://www.mma.gov.br/port/conama/res/res97/res23797.html>. Accessed on: January 31, 2010.

CURCIO, G. R.; LIMA, V. C.; GIAROLA, N. F. B. **Anthroposols**: proposed order. 1 approx. Colombo: EMBRAPA Florestas, 2004. 49 p. (EMBRAPA CNPF.

Documents, 101). CD-Rom.

CRAUL, P. J. A description of urban soil and their desired characteristics. Arboric. J., Urban, v. 11, n. 11, 1985. p. 330-339.DIAS JUNIOR, M.S. Compactaçao do solo. In: NOVAIS, R.F., ALVARES, V.H.; SCHAEFER, C.E.G.R. **Tópicos em ciência do solo**. Viçosa : SBCS, 2000. V.1, p.55-94.

DAINESE, R. C. **Remote sensing and geoprocessing applied to temporal land use and the comparison between non-supervised classification and visual analysis.** (Dissertation) Master's Degree in Agronomy - Faculdades de Ciências Agronômicas da UNESP, Botucatu, 2001. Available at: <http://www.athena.biblioteca.unesp.br/exlibris/bd/bla/33004064021P7/2001/daines e_rc_me_ botfca.pdf> Accessed on: January 15, 2011.

DALMOLIN, R. S. D.; AZEVEDO, A. C.; PEDRON, F. A. **Solos & Ambiente**, II Fórum. Santa Maria: Orium, 2006.

DERANI, C. **Direito ambiental económico**. 2. ed. Sao Paulo: Max Limonad, 2001.

DIAS JUNIOR, M.S. Soil compaction. In: NOVAIS, R.F.; ALVAREZ V., H.V. & CHAEFER, C.E.G.R. **Tópicos em ciência do solo**. Viçosa, Sociedade Brasileira de Ciência do Solo, 2000. v.1. p.55-94.

DRYZEK, J. Ecology and discursive democracy: beyond liberal capitalism and the administrative state. **Capitalism, Nature, Socialism**, Vol.3, No.2, 1992, pp.18-42

EMBRAPA. **Manual of soil analysis methods**. 2. ed. Rio de Janeiro: EMBRAPA Solos, 1997.

EMBRAPA. **Brazilian soil classification system**. Rio de Janeiro: EMBRAPA Solos, 1999.

EMBRAPA. **Brazilian soil classification system**. 2. ed. Rio de Janeiro: EMBRAPA Solos, 2006.

FARIAS, T. Principios gerais do direito ambiental. **Revista Primafacie**, ano 5, n. 9, jul./dez. 2006, pp. 126-148.

FERREIRA, Aurélio Buarque de Holanda. **Novo Aurélio Século XXI**: o dicionârio da lingua portuguesa. 3.ed. revised and expanded. - Rio de Janeiro: Nova Fronteira, 1999.

FLORIANO, E. P. Environmental Planning. **Caderno Didàtico**, n° 6, 1ª ed. Santa Rosa, 2004. 54 p.

FLORENZANO, T. G. **Satellite images for environmental studies**. Sao Paulo: Oficina de Textos, 2002.

FUJIMOTO, N. S. V. M. **Implicações ambientais na área metropolitana de Porto Alegre- RS:** um estudo geogràfico com ênfase na geomorfologia urbana. 2005. Available at: <http://www.geografia.ffch.usp.br/publicacoes/geousp/geousp12/geousp12_ninafujimoto.htm>. Accessed on November 25, 2008.

FRANCISCO FILHO, L. L. **Spatial distribution of violence in campinas: an analysis using geoprocessing**. Thesis. PhD in Geography. Department of Geography, Federal University of Rio de Janeiro, 2004. Available at: <http://www.fec.unicamp.br/~llfilho/Tese/Tese%20Final.pdf> Accessed on: January 15, 2011.

FREIRE, O. **Soils of tropical regions**. Botucatu: FEPAF (Fundaçao de Estudos e Pesquisas Agricolas e Florestais), 2006.

FREY, K. The political-democratic dimension in theories of sustainable development and its implications for local management. **Ambiente e Sociedade**. n.9. Campinas, Jul/Dec. 2001

GIELFF. S. PAC works in the municipality of Ourinhos/SP. Ourinhos, January 22, 2010. **Interview** given to Anahi Chimini Sobral.

GODOY, M. C. T. F. et al. Technogenic deposits in Presidente Prudente-SP: an example of an industrial district expansion area. In: CONGRESSO BRASILEIRO DE GEOGRAFIA FÌSICA APLICADA. XI., 2005. Sao Paulo. **Proceedings**...Sao Paulo: USP. p. 511-520. 1 CD. ROOM.

GOMES, D. C. H.; COELHO R. M.; GUADALUPE, G. W. Anthropogenic soils of the Anhumas stream basin: characterization, legend and environmental implications. In:. **Environmental recovery, participation and public power**: an experience in Campinas. AGRONOMIC INSTITUTE. 2nd phase report. Final report (May 2005 to June 2006). Available at <http://www.iac.sp.gov.br/projetoanhumas/relatorio.htm>. Accessed on: May 10, 2009.

GOULART, M.; CALLISTO, M. Bioindicators of water quality as a tool in environmental impact studies. **Revista da FAPAM**, year 2, n 1,

2003. Available at:

<http://www.icb.ufmg.br/big/beds/arquivos/goulartecallisto.pdf>. Accessed on February 1, 2010.

GUERRA, A. J. T.; CUNHA, S. B. (Org). **Geomorphology and the environment**. 5. ed. Rio de Janeiro: Bertrand Brasil, 2004.

GUERRA, A. J. T. Experiments and Monitoring in Soil Erosion. **Revista do Departamento de Geografia**, Universidade de Sao Paulo, n.16, p. 32-37, 2005.

TECHNOLOGICAL RESEARCH INSTITUTE OF THE STATE OF SÃO PAULO. **Geological map of the state of São Paulo**. Sao Paulo, 1981. Scale 1:500.000.

KIEHL, E. J. **Manual de edafologia**: relações solo planta. Sao Paulo: Ceres, 1979.

KASZNAR, i. K. Programa de Aceleraçao do crescimento (PAc). in: oLiVEiRA, F. B. de org. **Politica de Gestâo Pùblica Integrada**. Rio de Janeiro, Editora FGV, 2008.

LAMPARELLi, R. A. c.; RocHA, J. V. & BoRGHi, E. **Geoprocessing and precision agriculture.** Guaiba: Agropecuâria, 2001. 118 p.

LAYRARGUES, P. P. **From eco-development to sustainable development:**

evolution of a concept? Proposal, 25(71):5-10.1997.

LEPScH, i. F. **Formação e conservação do solo**. Sao Paulo: oficina de Textos, 2002.

MAcEDo, R. K. de. What environmental zoning is and what it is used for. **Texto Eletrônico**, 1998. Available at: <http://www.tede.ufsc.br/teses/PGEA0187.pdf>. Accessed on: February 1, 2010.

MAcHADo, P. A. L, **Direito ambiental brasileiro**. 14. ed. Sao Paulo: Malheiros, 2006.

MARTINS, M. J. **Solo urbano, solo rural**: efeitos da classificaçao no território ibérico. instituto de Dinâmica do Espaço, Universidade Nova de Lisboa, s.a.

Available at:

<http://www.apgeo.pt/files/docs/cD_X_coloquio_iberico_Geografia/pdfs/070.pdf> Accessed on April 13, 2009.

MARTiNS, M. S. et al. Geoprocessing techniques applied to study the environmental risk of visceral leishmaniasis in the urban area of Feira de Santana, Bahia.

Proceedings of the 13th Brazilian Symposium on Remote Sensing, Florianópolis, 2007.

Available at:

<http://marte.dpi.inpe.br/rep/dpi.inpe.br/sbsr@80/2006/11.16.00.40?

mirror=dpi.inpe.br/banon/2003/12.10.19.30.54&metadatarepository=dpi.inpe.br/sbs r@80/200 6/11.16.00.40.12> Accessed on: January 15, 2011.

MEDEIROS, J. S.; CÂMARA, G. Geoprocessing for environmental projects. In: CÂMARA, G.; DAVIS, C.; MONTEIRO, A. M. V. (ed.). **Introduçâo à ciência da geoinformaçâo**. Sao José dos Campos: INPE, 2001. Available at: <goto/sid.inpe.br/sergio/2004/04.19.15.08>. Accessed on: April 23, 2004.

MENDONÇA, F. de A. **Geografia e meio ambiente**. 8. ed., 1ª reprint - Sao Paulo: Contexto, 2007.

MILARÉ, E. **Direito do ambiente**. 3. ed. Sao Paulo: Revista dos Tribunais, 2004.

MIRRA, A. L. V. **Impacto Ambiental**. Sao Paulo: Oliveira Mendes, 1999.

MOTA, S. **Urbanization and the Environment**. ABES - Brazilian Association of Sanitary and Environmental Engineering. 1999. v. 1. 352 p.

MONTANO, M. **Water resources and environmental zoning: the case of the municipality of Sâo Carlos (SP)**. SP - 2002. Dissertation (Master's Degree in Hydraulics and Sanitation) Sao Carlos School of Engineering - USP. Available at: <sistemas3.usp.br/tycho/produocaacademica/eesc/shs/OC2.html>. Accessed on: January 28, 2010.

MOREIRA, A. C. M. L. **Megaprojects & the urban environment**: methodology for preparing a Neighborhood Impact Report. SP - 1997. Thesis (Doctorate) Faculty of Architecture and Urbanism - FAU, USP, Sao Paulo.

MOREIRA, M. A. **Fundamentals of remote sensing and application methodologies**. Sao José dos Campos: INPE, 2001.

MOURA, A. C. M. **Geoprocessamento na gestâo e planejamento urbano**. 2. ed. Belo Horizonte: Ed. da autor, 2005.

NEFUSSI, N.; LICCO, E. **Urban Soil and the Environment**. Available at : <http://www.mre.gov.br/cdbrasil/itamaraty/web/port/meioamb/mamburb/apresent/a present.ht m> Accessed on April 30, 2005.

NUNES, J. O. R. **Environmental planning**: comparative analysis of argissolos and latossolos using the profile homogenization method to obtain cover material for sanitary and controlled landfills for the region of Presidente Prudente and municipalities in the west of São Paulo. Scholarship Report for the Temporary Doctoral Absorption Program - Post-doc, 2003.

OLIVEIRA, J. B.; JACOMINE, P. K. T.; CAMARGO, M. N. **General soil classes of Brazil**: an auxiliary guide for their recognition. 2. ed. Jaboticabal: FUNEP, 1992.

OLIVEIRA, J. B. et al. **Pedological map of the state of Sao Paulo:** expanded legend. Campinas: EMBRAPA, 1999.

OLIVEIRA, A. M. S. et al. Technogen: records of man's geological action. In: SOUZA, C. R. G. et al. **Quaternary of Brazil**. Ribeirao Preto: Holos, p. 363-378. 2005.

OURINHOS. **Complementary Law** No. 499 of December 28, 2006. Available at: <http://ourinhos.prefeituramunicipal.net/ws/servicos/arquivos.php?cat=2>. Accessed on: January 15, 2010.

PHILLIP JR., A.; ALVES, A. C. **Curso interdisciplinar de direito ambiental**. Barueri: Manole, 2005.

PRADO FILHO, J. F. do; SOUZA, M. P. de. The environmental licensing of mining in the Iron Quadrangle of Minas Gerais - an analysis of the implementation of environmental control measures formulated in EIAS/RIMAS.

Rev. de Engenharia Sanitària Ambiental,

v. 9 - n. 4 - oct/dec 2004, 343-349. Available at: <libdigi.unicamp.br/document/?view=000433177>. Accessed on: January 28, 2010. .

PEDRON, F. A.; DALMOLIN, R. S. D. Characterization and management of urban soils. In:

Technical Report. Center for Rural Sciences, Federal University of Santa Maria, 2002.

PEDRON, F. A., et al. Urban soils. **Ciência Rural**. v.34. n.5. Santa Maria: Sept./Oct. 2004. Available at: <http://www.scielo.br/pdf/cr/v34n5/a53v34n5.pdf> Accessed on April 15, 2009.

PEDRON, F. A., et al. Soil survey and classification in urban areas: importance, limitations and applications. **R. Bras. Agrociência**, Pelotas, v. 13, n.2, p. 147-151, Apr/Jun, 2007. Available at: <http://www.ufpel.edu.br/faem/agrociencia/v13n2/artigo02.htm> Accessed on April 15, 2009.

PELOGGIA, A. **O homem e o ambiente geológico**: geologia, sociedade e ocupaçao urban no municipio de Sao Paulo. Sao Paulo: Xama, 1998.

PELOGGIA, A. A cidade, as vertentes e as vàrzeas: a transformação do relevo pela acção do homem no município de Sao Paulo. **Revista do Departamento de Geografia,** 2005, 24-31.

PERETTI, A.A. **History of the degraded area near Avenida Vitalina Marcusso**. Ourinhos: [s.n.], 2007. Personal communication.

PERUSI, M. C.; BERGAMASCHI, E. V. Elaboration and classification of a technogenic deposit profile resulting from land use and occupation in the Furnas stream - municipality of Ourinhos/SP. In: VII SIMPÒSIO NACIONAL DE GEOMORFOLOGIA SINAGEO E II ENCONTRO LATINO-AMERICANO DE GEOMORFOLOGIA. 2008, Belo Horizonte. **Proceedings**...Minas Gerais: UFMG, p. 124.

PRADO, H. **Solos do Brasil:** gênese, morfologia, classificaçao, levantamento, manejo agricola e geotécnico. 3.ed. Piracicaba: 2003.

OURINHOS CITY HALL. Available at: <http://www.ourinhos.sp.gov.br/> Accessed on: January 12, 2010.

PALMIERI, F.; LARACH, J. O. I. Pedology and geomorphology. In: GUERRA, A. J. T.; CUNHA, S. B. (org). **Geomorphology and the environment**. 5. ed. Rio de Janeiro: Bertrand Brasil, 2004. p. 59-122.

RAMOS, G. **Galeria pluvial de Ourinhos**. Ourinhos: [s.n.], 2007. Personal communication. REINERT, D. J.; REICHERT, J. M. **Physical properties of soil**. Santa Maria, 2006.

Available at:<http://w3.ufsm.br/msrs/textos%20dowload/apresentacoes%20aulas%20lectures/propriedades%20fisicas%20do%solo%20-%20reichert_reinert.pdf>. Accessed on: January 15, 2009.

RESENDE, M.; CURI, N.; REZENDE, S. B.; CÔRREA, G. F. **Pedologia:** base para distinçao de ambientes. 5. ed. Lavras: Editora UFLA, 2007.

RICARDO E. G.; RODRIGUES. A. M., HAAG. A. L. **The Growth Acceleration Program and public investment in the regions**. Available at:

<http://cedeplar.ufmg.br/seminarios/seminario_diamantina/2008/D08A061.pdf>. Accessed on: January 16, 2010.

ROCHA, V. L.; FORESTI, C. O uso do solo urbano e os problemas ambientais da cidade de Iguape-SP - **Anais IX Simpósio Brasileiro de Sensoriamento Remoto**. Santos: INPE, 1998.

RODRIGUES, A. M. **Desenvolvimento Sustentâvel:** dos conflitos de classes para o conflito de gerações. In: Panorama of Brazilian Geography. v.2. Annablume, 2006.

ROSS, J. L. S.; MOROZ, I. C. **Geomorphological map of the state of Sao Paulo**. 1997. Scale 1:50.000.

ROSSATO, M. S., SUERTEGARAY, D. **Rethinking the time of nature in transformation**. Agora (UNISC), Santa Cruz, v.6, n. 2, p. 93-98, 2000. Available at: <http://www.ufrgs.br/nega/Textos%20e%20artigos/Repensando%20o%20tempo%2 0da%20n ature%20em%20transforma%E7%E3o.pdf>. Accessed on January 16, 2010.

SALOMÂO, F. X. T. Control and prevention of erosion processes. In: GUERRA, A. J. T.; SILVA, A. S.; BOTELHO, R. G. M. (org.). **Erosion and soil conservation:** concepts, themes and applications. 2. ed. Rio de Janeiro: Bertrand Brasil, 2005.

SAMPAIO, E. **Basics of Remote Sensing and Photointerpretation in Pedology**. Department of Geosciences. University of Évora. Portugal, 2007.

SANTOS, R. F. dos. **Environmental planning**: theory and practice. Sao Paulo: Oficina de textos, 2004.

SANTOS, R, D., et al. **Manual de descriçao e coleta de solo no campo**. 5. ed. Viçosa: Brazilian Society of Soil Science, 2005.

SANTOS JUNIOR, J. B. **Residential urban soils of the Jardim Paulista neighborhood, Campina Grande do Sul (PR)**. Federal University of Paranà: Curitiba, 2008. Dissertation.

Available at:

<http://dspace.c3sl.ufpr.br/dspace/bitstream/1884/17209/1/Disserta%C3%A7%C3%

A3o%20J aime.pdf> Accessed on May 30, 2009.

SILVA, I. F.; MIELNICZUK, J. Action of the plant root system on the formation and stabilization of soil aggregates. **Brazilian Journal of Soil Science**.

Campinas, v. 21, n. 1, p. 113-117, jan./mar. 1997.

SILVA, A. M., et al. **Erosion and hydrosedimentology in river basins**. 1. ed. Sao Carlos: RIMA, 2003-2004.

SILVA, J. A. da. **Constitutional Environmental Law**. Sao Paulo: Malheiros, 1994

SILVA, J. dos S. V. **Multivariate analysis in zoning for environmental planning: Case study: Upper Taquari River basin - MS/MT**. SP - 2003. Thesis (Doctorate in Agricultural Engineering) Faculty of Agricultural Engineering - FEA, UNICAMP, Campinas.

SOUZA, L. C. de, O estudo prévio de impacto ambiental como instrumento de protecção ambiental. **Revista da Faculdade de Direito Padre Anchieta.** ano IV. n. 7. Novembro/2003.

SUERTEGARAY, D. M. A. Geomorphology: new concepts and approaches. In: VII BRAZILIAN SYMPOSIUM OF GEOGRAPHY AND I AMERICAN FORUM OF GEOGRAPHY

APPLIED PHYSICS. VII and I, 1997. Curitiba. **Proceedings**...Curitiba. p. 24-29.

STRECK, E. V. et al. **Soils of Rio Grande do Sul**. 2. ed. Porto Alegre: EMATER/RSASCAR, 2008.

VEIGA, T. C.; XAVIER-DA-SILVA, J. Geoprocessing applied to the identification of potential areas for tourist activities: the case of the municipality of Macaé - RJ. In: XAVIER- DASILVA, J.; ZAIDANR. T. (organizers). **Geoprocessing and environmental analysis**: applications. Rio de Janeiro: Bertrand Brasil, 2004.

VEIRA, L. S. **Manual da Ciência do Solo**: com ênfase aos Solos Tropicais. 2. ed. Sao Paulo: Ed. Agronomica Ceres, 1988.

XAVIER-DA-SILVA, J.; ZAIDAN, R. T. Geoprocessing applied to the zoning of areas in need of protection: the case of Ibitipoca State Park - MG. In: XAVIER-DA-SILVA, J.; ZAIDANR. T. (organizers). **Geoprocessing and environmental analysis**: applications. Rio de Janeiro: Bertrand Brasil, 2004.

ZACHARIAS, A. A. **The graphic representation of landscape units in environmental zoning: a case study in the municipality of Ourinhos** - SP.

2006. 200 f. Thesis (Doctorate in Geosciences), Institute of Geosciences and Exact Sciences - IGCE, UNESP, Rio Claro.

Printed by Books on Demand GmbH, Norderstedt / Germany